# HIGH TEMPERATURE CREEP–FATIGUE

# CURRENT JAPANESE MATERIALS RESEARCH

*Volume 1*
Current Research on Fatigue Cracks
T. TANAKA, M. JONO and K. KOMAI

*Volume 2*
Statistical Research on Fatigue and Fracture
T. TANAKA, S. NISHIJIMA and M. ICHIKAWA

# HIGH TEMPERATURE CREEP-FATIGUE

*Edited by*

## RYUICHI OHTANI

*Kyoto University, Japan*

## MASATERU OHNAMI

*Ritsumeikan University, Japan*

## TATSUO INOUE

*Kyoto University, Japan*

*Current Japanese Materials Research — Vol. 3*

ELSEVIER APPLIED SCIENCE
LONDON and NEW YORK

ELSEVIER APPLIED SCIENCE PUBLISHERS LTD
Crown House, Linton Road, Barking, Essex IG11 8JU, England

*Sole Distributor in the USA and Canada*
ELSEVIER SCIENCE PUBLISHING CO., INC.
52 Vanderbilt Avenue, New York, NY 10017, USA

WITH 25 TABLES AND 210 ILLUSTRATIONS

© ELSEVIER APPLIED SCIENCE PUBLISHERS LTD and
THE SOCIETY OF MATERIALS SCIENCE, JAPAN 1988

**British Library Cataloguing in Publication Data**

Ohtani, Ryuichi
    High temperature creep–fatigue. —
    — (Current Japanese materials research; V.3)
    1. Materials — Fatigue    2. Materials
    at high temperatures
    I. Title      II. Ohnami, Masateru
    III. Inoue, Tatsuo      IV. Series
    620.1'123      TA418.38

    ISBN 1-85166-135-2

**Library of Congress Cataloging-in-Publication Data**

High temperature creep–fatigue/edited by Ryuichi Ohtani, Masateru
    Ohnami, Tatsuo Inoue.
        p. cm. — (Current Japanese materials research; vol. 3)
        Bibliography: p.
        Includes index.
        ISBN 1–85166–135–2
        1. Materials — Creep.   2. Materials — Fatigue.   3. Materials at high
    temperatures.   I. Ohtani, Ryuichi, 1938–   .   II. Ohnami, Masateru,
    1931–   .   III. Inoue, Tatsuo, 1939–   .   IV. Series.
    TA418.22.H54 1988
    620.1'1233--dc19                                                      87–21312

**Special regulations for readers in the USA**

Phototypesetting by Tech-Set, Gateshead, Tyne & Wear.
Printed in Great Britain at the University Press, Cambridge.

# Foreword

The Current Japanese Materials Research (CJMR) series is a new publication edited by the Society of Materials Science, Japan, and published by Elsevier Applied Science, UK, aiming at the overseas circulation of current Japanese achievement in the field of materials science and technology. This third volume of the series deals with *High Temperature Creep-Fatigue* and follows the second volume on *Statistical Research on Fatigue and Fracture*. All the papers have been selected to present the most important and substantial results obtained by the authors, in order to help readers to understand the combined micro and macro mechanical approaches concerning high temperature creep-fatigue which have been studied in Japan.

Although many international meetings are held every year in various specialized fields, it cannot be denied that most research results in Japan are published only in Japanese and tend therefore to be confined to the domestic audience. The publication of the CJMR series is an attempt to offer these results to colleagues abroad and thereby encourage the exchange of knowledge between us. I hope that our efforts will interest engineers and scientists in different countries and may contribute to the progress of materials science and technology throughout the world.

KIYOSHI OKADA
President, Society of Materials Science, Japan

# Preface

An increase in the efficiency of power generating units, in the rate of industrial chemical processes, or in the speed of many kinds of aircraft is closely related to an increase in temperature. Fatigue at high temperature is encountered in these engineering structures in such ways that high frequency vibration as a consequence of rotation or fluid motion brings on high cycle fatigue failure, and relatively low frequency strain cycling as a consequence of start and stop operations generates low cycle fatigue failure. The origin of interest in the subject of fatigue at high temperature is thus far from purely academic. The design objective of structures is to provide reliable performance with a minimum of break-downs and minimum repair costs. However, before the development of design methods can be put into practice, it is necessary that the structural materials are thoroughly characterized with respect to their resistance to stress, temperature and the environment. Although technological advances are greatly dependent on the development of new heat resisting materials, such a design can only be made reliable after the mechanical behavior of the materials has been quantified.

Experience in the field of technology in Japan has also shown that fatigue may often be the controlling factor for the failure of components in high temperature service. Such a state of affairs has offered a challenge to researchers and design engineers in Japan with the topic of creep–fatigue interaction. Our fundamental understanding of the phenomena of high temperature creep–fatigue in engineering materials has advanced significantly in recent years, but the mechanics and the mechanisms are still obscure, because high temperatures introduce a number of complications to deformation and fracture in the creep–

fatigue process. The Committee on High Temperature Strength of Materials in the Society of Materials Science, Japan, (JSMS), which is one of the most active research and communication groups with about 100 regular members, of whom half belong to industrial research and development institutes, has been playing an important part in motivating a combined science and technology approach to creep-fatigue. The present volume is intended to provide an overview, as well as current research findings on this subject. Although the number of papers contained in this volume is limited, readers are asked to interpret the contents as an interim report *en route* to our goal.

The volume begins with chapters on the inelastic constitutive relationship under creep–plasticity interaction conditions. Particular attention is given to the theoretical consideration of complicated deformation behavior in thermally activated processes. This introduces the need to establish the general formulation of stress–strain–time–temperature interactions not only to calculate varying-load creep deformation or thermal ratchetting, but also to evaluate failure life under creep–fatigue interaction conditions. Moreover, to give fundamental ideas on the global material failure at high temperature, a continuum damage theory is presented with special emphasis on the phenomenological representation of defects or cracks.

Advanced aspects of creep and fatigue crack growth are treated in the following section, these being comprehensive reviews of work on fracture mechanics that contribute to an understanding of the substance of creep–fatigue interaction. The difference of fracture mode and damage rule among different kinds of steel are also indicated by analyzing the resultant data from careful tests. The influence of environmental air on low cycle fatigue failure life is discussed, based on the strain range partitioning approach.

The last section is devoted to life prediction, and to the still very open research area of the methodological approach to damage evaluation and life or remaining life estimation. Some new concepts are presented, which are applied to combined stress fatigue, thermal–mechanical fatigue and low cycle fatigue in welded joints. In this section are included reviews of current approaches to the creep–fatigue life prediction of high temperature nuclear reactor components and steam turbine rotors. The applicability of a non-destructive assessment of material degradation to the remaining life prediction of steam turbine rotors is also discussed.

We sincerely hope that the work done by the authors presented in this volume will be of lasting value. It is also hoped that discussions will follow between the authors and those who are interested in the authors' research to deepen our mutual understanding and to reach our common goals. On behalf of the Committee on High Temperature Strength of Materials, JSMS, we would like to thank all those who contributed a significant amount of their time and dedication to the publication of this volume. It is only through their efforts that this publication has come into being.

RYUICHI OHTANI
MASATERU OHNAMI
TATSUO INOUE

# Contents

## Life Prediction

# List of Contributors

KAZUNARI FUJIYAMA
*Heavy Apparatus Engineering Laboratory, Toshiba Corporation, 1-9 Suehiro-cho, Tsurumi-ku, Tokohama City, Japan*

HIROSHI HATTORI
*Research Institute, Ishikawajima-Harima Heavy Industries Co. Ltd., 3-1-5 Toyosu Kotoku, Tokyo 135, Japan*

TOSHIHIDE IGARI
*Nagasaki Research and Development Center, Mitsubishi Heavy Industries, Ltd., 1-1 Akunoura-machi, Nagasaki 850-91, Japan*

SHOJI IMATANI
*Department of Mechanical Engineering, Kyoto University, Sakyo-ku, Kyoto 606, Japan*

TATSUO INOUE
*Department of Mechanical Engineering, Kyoto University, Sakyo-ku, Kyoto 606, Japan*

YOSHIYASU ITOH
*Heavy Apparatus Engineering Laboratory, Toshiba Corporation, Turumi-ku, Yokohama, Japan*

OSAMA KANEMARU
*Creep Testing Division, National Research Institute for Metals, 2-3-12 Nakameguro, Meguro-ku, Tokyo, Japan*

LIST OF CONTRIBUTORS

KAZUSHIGE KIMURA

*Heavy Apparatus Engineering Laboratory, Toshiba Corporation, 1-9 Suehiro-cho, Tsurumi-ku, Tokohama City, Japan*

MASAKI KITAGAWA

*Research Institute, Ishikawajima-Harima Heavy Industries Co. Ltd., 3-1-15 Toyosu Kotoku, Tokyo 135, Japan*

TAKAYUKI KITAMURA

*Department of Engineering Science, Faculty of Engineering, Kyoto University, Yoshida, Sakyo-ku, Kyoto 606, Japan*

KIYOSHI KUBO

*Creep Testing Division, National Research Institute for Metals, 2-3-12 Nakameguro, Meguro-ku, Tokyo, Japan*

SHIRO KUBO

*Department of Mechanical Engineering, Osaka University, 2-1 Yamadaoka, Suita, Osaka 565, Japan*

KAZUO KUWABARA

*Nuclear Engineering Department, Komae Research Laboratory, Central Research Institute of Electric Power Industry, 2-11-1 Iwato-kita, Komae-shi, Tokyo 201, Japan*

SUMIO MURAKAMI

*Department of Mechanical Engineering, Nagoya University, Furo-cho, Chikusa-ku, Nagoya 464, Japan*

MASAMITSU MURAMATSU

*Heavy Apparatus Engineering Laboratory, Toshiba Corporation, 1-9 Suehiro-cho, Tsurumi-ku, Tokohama City, Japan*

YOSHIHARU MUTOH

*Department of Mechanical Engineering, Technological University of Nagaoka, Tomioka, Nagaoka, Japan*

SEIICHI NISHINO

*Graduate School of Ritsumeikan University, 56-1 Tojiin Kita-machi, Kita-ku, Kyoto 603, Japan*

AKITO NITTA
*Nuclear Engineering Department, Komae Research Laboratory, Central Research Institute of Electric Power Industry, 2-11-1 Iwato-kita, Komae-shi, Tokyo 201, Japan*

ISAMU NONAKA
*Research Institute, Ishikawajima-Harima Heavy Industries Co. Ltd., 3-1-15 Toyosu Kotoku, Tokyo 135, Japan*

KIYOTSUGU OHJI
*Department of Mechanical Engineering, Osaka University, 2-1 Yamadaoka, Suita, Osaka 565, Japan*

MASATERU OHNAMI
*Department of Mechanical Engineering, Ritsumeikan University, 56-1 Tojiin Kita-machi, Kita-ku, Kyoto 603, Japan*

NOBUTADA OHNO
*Department of Energy Engineering, Toyohashi University of Technology, Tempaku-cho, Toyahashi 440, Japan*

RYUCHI OHTANI
*Department of Engineering Science, Faculty of Engineering, Kyoto University, Yoshida, Sakyo-ku, Kyoto 606, Japan*

AKIRA OHTOMO
*Research Institute, Ishikawajima-Harima Heavy Industries Co. Ltd., 3-1-15 Toyosu Kotoku, Tokyo 135, Japan*

MASAKAZU OKAZAKI
*Department of Mechanical Engineering, Technological University of Nagaoka, Tomioka, Nagaoka, Japan*

MASAO SAKANE
*Department of Mechanical Engineering, Ritsumeikan University, 56-1 Tojiin Kita-machi, Kita-ku, Kyoto 603, Japan*

KATSUYA SETOGUCHI
*Nagasaki Research and Development Center, Mitsubishi Heavy Industries, Ltd., 1-1 Akunoura-machi, Nagasaki 850-91, Japan*

CHIAKI TANAKA

> *Creep Testing Division, National Research Institute for Metals, 2-3-12 Nakameguro, Meguro-ku, Tokyo, Japan*

KATUYUKI TOKIMASA

> *Technical Research Laboratory, Sumitomo Metal Industries, Ltd., Amagasaki, Japan*

KOICHI YAGI

> *Creep Testing Division, National Research Institute for Metals, 2-3-12 Nakameguro, Meguro-ku, Tokyo, Japan*

MASAFUMI YAMAUCHI

> *Nagasaki Research and Development Center, Mitsubishi Heavy Industries, Ltd., 1-1 Akunoura-machi, Nagasaki 850-91, Japan*

# Inelastic Constitutive Relationship of High Temperature Materials under Creep-Plasticity Interaction Conditions

TATSUO INOUE and SHOJI IMATANI

*Department of Mechanical Engineering, Kyoto University, Sakyo-ku, Kyoto 606, Japan*

## ABSTRACT

A unified type of constitutive model is developed to describe creep–plasticity interaction conditions. General discussion based on excess stress theory is stated in the first part of the chapter, and applied to the special case of an inelastic constitutive model capable of expressing the creep–plasticity interaction behavior under complex loading paths. Experimental verifications of the theory as well as other models already developed are carried out on a tension–compression and cyclic torsion testing machine. A brief description of the experimental system by use of a personal computer is given. The use of this inelastic constitutive relationship for high temperature materials is also examined to test the applicability of the theory under the complicated loading history.

## INTRODUCTION

An appropriate representation of the inelastic constitutive relationship for high temperature materials has been required to facilitate the design of high temperature machine components. This representation should also provide the fundamental information for life prediction under the creep–fatigue regime, because it is well known that materials exhibit a characteristic rate dependence under such an environment originating from the interaction between creep and plasticity.

In order to describe an accurate stress–strain relationship even under complicated loading paths, several kinds of constitutive equations [1–11] have been proposed, which can be roughly classified into two types: one is the conventional superposition model [1, 2] comprising

1

constitutive plastic and creep equations, and the other is the unified model that is based on certain constitutive laws, such as the excess (over-) stress theory [3, 4], the endochronic theory [5] and others [6–11]. The latter type of model seemed to reflect the real material behavior, since it is hard to distinguish between creep and plastic strains in the physical deformation mechanism. Theories on the bounding surface [12, 13] or additional hardening surface in strain space [14, 15] have also been developed in order to describe the nonlinear cyclic hardening behavior of materials.

The purpose of this paper is to provide a unified inelastic constitutive equation to satisfy the creep–plasticity interaction condition, and the validity and applicability of the model will be examined for high temperature steels. An inelastic constitutive model is first developed that is based on the excess stress hypothesis originally presented by Perzyna [3] and modified to fit the real material behavior. Evolution equations for internal state variables are determined in order that the model can describe the rate-dependent stress–strain relation including creep. Subsequently, an experimental verification of the model is carried out by using an electrohydraulic testing machine capable of reproducing the combined stress state of tension and torsion. Material responses under the following three series of stress and/or strain patterns are examined:

(1) monotonic tension with creep prestrain and its inverse pattern;
(2) the change in strain trajectory during proportional straining;
(3) mechanical ratcheting with varying stress.

The inelastic behavior obtained in these experiments will be examined to test the validity of the model as well as further results concerning anisotropy during the progressive deformation.

## INELASTIC CONSTITUTIVE EQUATION FOR CREEP–PLASTICITY INTERACTION

A unified inelastic constitutive model is proposed to account for the time-dependent material behavior by applying the excess stress theory originally developed by Perzyna and modified by several researchers.

### An Inelastic Constitutive Model Based on Excess Stress Theory

The excess stress theory [3] assumes that the viscoplastic strain rate $\dot{\varepsilon}^i$ is induced by the overstress exceeding the static yield stress $r$, and that the magnitude is proportional to the degree of the overstress. These assumptions are schematically illustrated in Fig. 1, where the direction of the inelastic strain rate is normal to the limit surface, so that it may be generally expressed as

$$\dot{\varepsilon}^i = \eta \Phi \langle F(\sigma) - r \rangle \frac{\partial F}{\partial \sigma} \quad \Phi \langle X \rangle = \begin{cases} \Phi(X), & \text{if } X \geqslant 0 \\ 0, & \text{if } X < 0 \end{cases} \quad (1)$$

Here, $\eta$ denotes the viscosity and $F(\sigma)$ represents a yield function. Equation (1) covers a wide range of viscoplastic constitutive laws, since we can choose the proper values of the parameters $\eta$ and $r$.

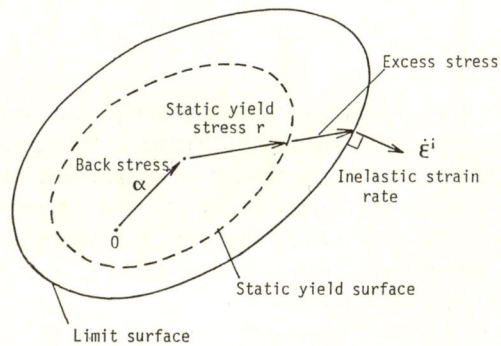

FIG. 1. General concept of the excess stress hypothesis in stress space.

When the kinematic hardening rule is adopted and the variation of viscosity during the inelastic deformation is considered, Eqn. (1) may be modified to

$$\dot{\varepsilon}^i = \Phi \left\langle \frac{F(\sigma - \alpha) - r}{D} \right\rangle \frac{\partial F}{\partial \sigma} \quad (2)$$

where $\alpha$ indicates the kinematic back stress and $D$ is the isotropic drag stress. Employing von Mises type of yield criterion, Eqn. (2) can be solved to yield

$$\sigma = |D\Phi^{-1}(\dot{\varepsilon}^i) + r| \, \text{sgn}(\dot{\varepsilon}^i) + \alpha \quad (3)$$

under the uniaxial state, which means that this constitutive model is composed of a parallel combination of viscous stress $D\Phi^{-1}(\bullet)$, static yield stress $r$ and kinematic back stress $\alpha$ as illustrated in Fig. 2.

This has evolved the inelastic strain, and we can now formulate the

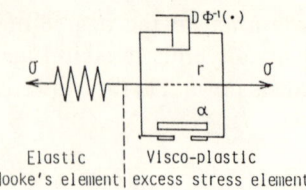

FIG. 2. Structure of the unified constitutive model proposed.

evolution equations for the internal variables appearing in Eqn. (3). According to Bailey-Orowan's form, the rate of back stress is expressed as

$$\dot{\alpha}' = \frac{2}{3} C_1 \dot{\varepsilon}^i - \frac{C_1}{C_2} \alpha' \tag{4}$$

in which $C_1$ and $C_2$ are non-negative material functions generally expressed by the current value of the stress, internal variables and temperature. The parameter $C_1$ represents the hardening due to inelastic deformation, while $C_2$ is considered to be the recovery arising from the thermal activation process. It should be noticed that the structure of the back stress is composed of a linear combination of the (static) slip element and the viscous dashpot element.

The static yield stress $r$ is essentially rate independent, and therefore the rate may be expressed as

$$\dot{r} = k\dot{\kappa}_s \tag{5}$$

in which $\dot{\kappa}_s$ indicates a certain kind of static hardening parameter. Since we are dealing with a unified constitutive model, it may strictly be impossible to define such a suitable rate-independent (static) parameter. However, the back stress is composed of the linear combination of the slip element due to hardening and the viscous dashpot element indicating recovery, so that this static parameter can be considered as the path of the slip element. For instance, $\dot{\kappa}_s$ can be expressed as

$$\dot{\kappa}_s = \left\{ \frac{2}{3} \mathrm{tr} \left( \dot{\varepsilon}^i - \frac{3}{2} \frac{1}{C_2} \alpha' \right)^2 \right\}^{1/2} \tag{6}$$

referring to the evolution equation for back stress. Equation (6) coincides with the equivalent plastic strain rate when $C_2$ approaches infinity, whereas no hardening occurs under an equilibrium state of back stress like steady state creep.

Since the drag stress $D$ is rate dependent, on the contrary, the evolution equation can be reasonably assumed to be of a similar form to Eqn. (4) like

$$\dot{D} = p\,(\dot{\kappa}_v - qD) \tag{7}$$

Here, the viscous parameter $\dot{\kappa}_v$, incorporating the dissipation rate, can be determined as

$$\dot{\kappa}_v = \langle F(\sigma - \alpha) - r \rangle\,\bar{\varepsilon}^i. \tag{8}$$

All the parameters appearing in this model can be identified by conventional test data of monotonic tension, creep and cyclic straining using the next procedure.

The function $\Phi(\cdot)$ is of the same form as a creep equation whose parameters can be determined by a creep curve. The uniaxial stress-strain diagram plays an important role in identifying the parameters $C_1$ and $C_2$ in the back stress (see Eqn. (4)). Finally, the cyclic saturation and an additional simulation for the creep curve give a suitable set of parameters for the isotropic variables $r$ and $D$.

## Other Constitutive Models Employed in the Simulation

The following four types of existing constitutive models were also employed in the simulation of inelastic material behavior to compare with the experimental results:

(1) the superposition model, in which the inelastic strain rate is composed of a time-independent plastic strain and a rate-dependent creep strain in the classical form;

(2) the modified Mróz model, in which a modified creep equation [16] was superimposed on the original form by Kujawski and Mróz [6] in order to describe the creep deformation below yield stress;

(3) the Chaboche model [4] is also based on the excess stress hypothesis with internal state variables of kinematic back stress and static yield stress;

(4) the Miller model [7], in which the kinematic back stress and isotropic drag stress are defined by applying the creep equation by Garofalo.

A summary of the fundamental equations is presented in Table 1, where the symbols follow the nomenclature in customary usage. All the material parameters in each constitutive model can be determined by

## TABLE 1
### Fundamental equations of each constitutive model

| Superposition model | Mróz model |
|---|---|
| $F(\mathbf{s}, \boldsymbol{\alpha}', \kappa) = \frac{1}{2}\text{tr}(\mathbf{s}-\boldsymbol{\alpha}')^2 - \frac{1}{3}\bar{\sigma}(\kappa)^2 = 0$ | $F(\mathbf{s}, \boldsymbol{\alpha}', \boldsymbol{\beta}', \bar{\varepsilon}^p, \bar{\varepsilon}^p_c) = \frac{1}{2}\text{tr}(\mathbf{s}-\boldsymbol{\alpha}'-\boldsymbol{\beta}')^2 - \frac{1}{3}\bar{\sigma}(\bar{\varepsilon}^p)^2 = 0$ |
| $\dot{\boldsymbol{\alpha}}' = \frac{2}{3}C\dot{\boldsymbol{\varepsilon}}^p \quad C = \text{constant}$ | $\dot{\boldsymbol{\alpha}}' = \frac{2}{3}C(\bar{\varepsilon}^p_c)\dot{\boldsymbol{\varepsilon}}^p - D(\bar{\varepsilon}^p)\bar{\varepsilon}^p\frac{2}{3}\boldsymbol{\varepsilon}^p$ |
| | $\dot{\boldsymbol{\beta}}' = \frac{2}{3}C_2\dot{\boldsymbol{\varepsilon}}^p - \frac{C_2}{C_2}\boldsymbol{\beta}'$ |
| | $\bar{\varepsilon}^p_c = \bar{\varepsilon}^p_m + \bar{\varepsilon}^p + \bar{\varepsilon}^p_{k-1}$ |
| $\bar{\sigma} = \bar{\sigma}_\infty\{1 - l\exp(-\beta\bar{\varepsilon}^p)\}$ | |
| $\dot{\boldsymbol{\varepsilon}}^p = \frac{9}{4\bar{\sigma}^2}\frac{\{\text{tr}(\mathbf{s}-\boldsymbol{\alpha}')\dot{\mathbf{s}}\}}{(C+\bar{\sigma}')}(\mathbf{s}-\boldsymbol{\alpha}')$ | $\dot{\boldsymbol{\varepsilon}}^p = \frac{9}{4\bar{\sigma}^2}\frac{\{\text{tr}(\mathbf{s}-\boldsymbol{\alpha}'-\boldsymbol{\beta}')\dot{\mathbf{s}}\} + C_2/C_3\{\text{tr}(\mathbf{s}-\boldsymbol{\alpha}'-\boldsymbol{\beta}')\boldsymbol{\beta}'\}}{C+C_2-D/\bar{\sigma}\{\text{tr}(\mathbf{s}-\boldsymbol{\alpha}'-\boldsymbol{\beta}')\boldsymbol{\varepsilon}^p\} + \bar{\sigma}'}(\mathbf{s}-\boldsymbol{\alpha}'-\boldsymbol{\beta}')$ |
| $\dot{\boldsymbol{\varepsilon}}^c = \frac{3}{2}mA^{1/m}\bar{s}^{-(n-m)/m}\bar{\varepsilon}^{-c(m-1)/m}\mathbf{s}$ | $\dot{\boldsymbol{\varepsilon}}^c = \frac{3}{2}mA^{1/m}\bar{\xi}^{(n-m)/m}\bar{\zeta}^{(m-1)/m}(\mathbf{s}-\boldsymbol{\beta}')$ |
| | $\bar{\xi} = \{\frac{3}{2}\text{tr}(\mathbf{s}-\boldsymbol{\beta}')^2\}^{1/2}$ |
| | $\bar{\zeta} = \bar{\varepsilon}^c_c + \eta(\bar{\varepsilon}^p, \bar{\varepsilon}^p_c)$ |

*Chaboche model*

$$\dot{\varepsilon}^i = \frac{3}{2}\left\langle \frac{\bar{\xi} - R - r}{D}\right\rangle^n \frac{\xi}{\|\xi\|}$$

$$\xi = s - \alpha'$$

$$\bar{\xi} = \{\tfrac{3}{2}\mathrm{tr}(s-\alpha')^2\}^{1/2}$$

$$\langle X\rangle^n = \begin{cases} X^m, & \text{if } X \geq 0 \\ 0, & \text{if } X < 0\end{cases}$$

$$\alpha = \alpha_1 + \alpha_2$$

$$\dot{\alpha}'_1 = C_1(A_1\tfrac{2}{3}\dot{\varepsilon}^i - \bar{\dot{\varepsilon}}^i\alpha_1) - h\bar{\alpha}_1'^{\,m-1}\alpha_1'$$

$$\dot{\alpha}'_2 = \tfrac{2}{3}C_2\dot{\varepsilon}^i$$

$$\dot{R} = b(Q - R)\bar{\dot{\varepsilon}}^i - p|R|^q$$

*Miller model*

$$\dot{\varepsilon}^i = B\,\frac{3}{2}\left\{\sinh\left(\frac{\bar{\xi}}{\sqrt{D_s + D}}\right)^{1.5}\right\}^n \frac{\xi}{\|\xi\|}$$

$$\xi = s - \alpha'$$

$$\bar{\xi} = \{\tfrac{3}{2}\mathrm{tr}(s-\alpha')^2\}^{1/2}$$

$$\dot{\alpha}' = \tfrac{2}{3}H_1\dot{\varepsilon}^i - H_1 B\{\sinh(A_1\bar{\alpha}')\}^n\frac{\alpha'}{\bar{\alpha}}$$

$$\dot{D} = H_2\bar{\dot{\varepsilon}}^i\left(C_2 + \bar{\alpha}' - \frac{A_2}{A_1}D^{1.5}\right) - H_2 C_2 B\{\sinh(A_2 D^{1.5})\}^n$$

conventional test data for monotonic tension, creep and cyclic straining in the same manner with the proposed model, and can be almost independently picked up from the tests.

## EXPERIMENTAL SYSTEM AND THE PROCEDURE

Two kinds of high temperature materials, SUS304 stainless steel and $2\frac{1}{4}$ Cr–1Mo steel, were prepared for the experiment. Thin-walled tubular specimens were employed to realize the combined stress state, the shape being shown in Fig. 3. The SUS304 steel examined was the material solid-solution treated from 1100 °C, and the $2\frac{1}{4}$ Cr–1Mo steel was normalized for 60 min at 930 °C with subsequent tempering at 690 °C for 130 min. The experiments were performed at 650 °C for the 304 steel and at 600 °C for the Cr–Mo steel, the samples being kept free from stress at the setting levels of temperature for a while before the tests started.

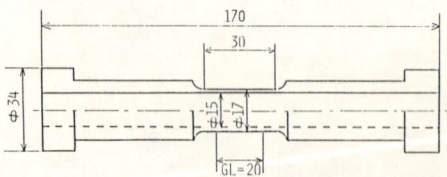

FIG. 3. Shape and dimensions of the specimens (dimensions given in millimeters).

A system to control the electrohydraulic testing machine that was capable of combined tension–compression and torsion was developed using a personal computer, a block diagram being presented in Fig. 4 [16]. Several parts of the subsystem (interfaces) in this figure were developed in the authors' laboratory:

(1) four sets of A/D converters to read analogue data and to convert them to digital form for the computer;
(2) two sets of D/A converters to input the setting signals to the machine;
(3) analogue mode switches to change the control signal;
(4) data memory module with 128K bytes;
(5) time clock generator for the interrupting process;
(6) interface for the choice of subsystem;

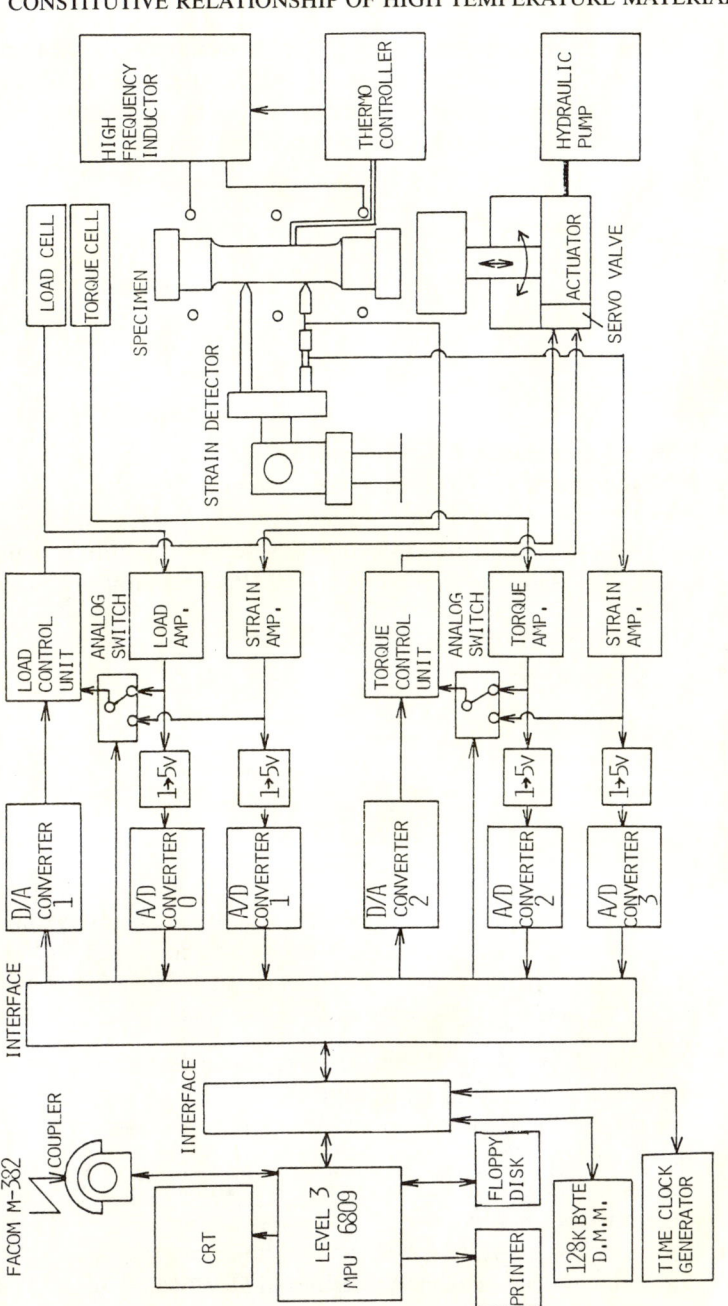

FIG. 4. Control system for the testing apparatus.

(7) interface to back up bus-line between the external modules and the computer;

(8) voltage amplifier to adapt the voltage between the components;

(9) strain detector for axial displacement and twisting angle.

## COMPARISON BETWEEN SIMULATED RESULTS AND EXPERIMENTAL DATA

The material parameters were first identified by conventional experimental data, and three types of loading path were simulated to examine the validity of the constitutive models.

### Creep–Plasticity Interaction Behavior under Monotonic Loading

Figure 5 shows the stress–strain diagram with an inserted creep strain for SUS304 steel, in which subsequent plastic deformation was investigated to determine the recovery effect as well as hardening during the creep process. The steady stress of 176 MPa was imposed up to

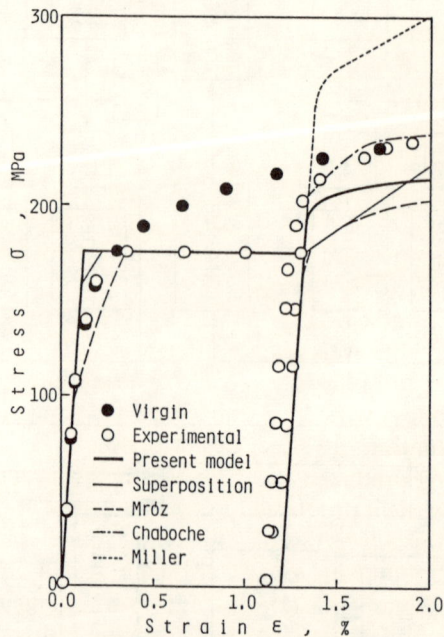

FIG. 5. Stress–strain diagram of creep-prestrained SUS304 steel at 650°C.

$\varepsilon = 1.2\%$ as the prior creep, and was followed by rapid straining under a strain rate of $\dot{\varepsilon} = 0.1\% \text{ s}^{-1}$. The experimental data are plotted as open circles, while solid circles indicate the monotonic tension at $\dot{\varepsilon} = 0.1\% \text{ s}^{-1}$ for comparison. As can be seen from the figure, a hardening effect on subsequent plastic deformation was detected even by the prior creep. Figure 6 presents creep curves under a stress of $\sigma = 120$ MPa with and without plastic prestraining for $2\frac{1}{4}$Cr–1Mo steel, in which a plastic prestrain was given at $0.5\% \text{ s}^{-1}$ up to $0.5\%$ as indicated by open circles, solid circles being the result of a simple creep test. Remarkable

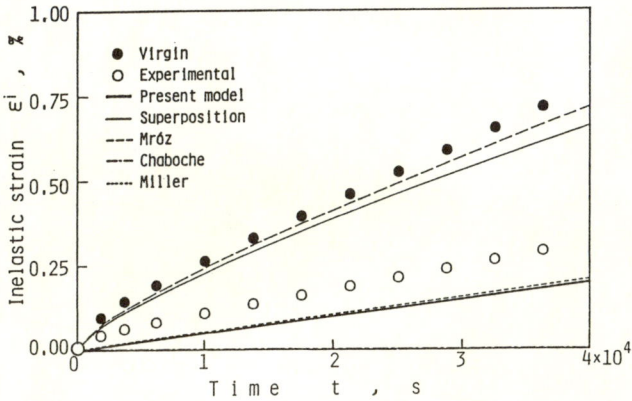

FIG. 6. Creep curves at the stress level of $\sigma = 120$ MPa for $2\frac{1}{4}$Cr–1Mo steel at 600 °C with 0·5% plastic strain.

hardening can be observed in the creep curve because of the prior plastic deformation.

The calculated results are represented by lines in the figure, and the proposed model qualitatively gives sufficient results for the hardening effects due to both prior plastic and creep deformations, although there is some discrepancy with the experimental results. This model succeeded in evaluating the effect of the preloadings. The Chaboche model seems to have predicted a reasonable stress–strain relation as can be seen in Fig. 5, while it produced too small a degree of creep followed by plastic prestraining (see Fig. 6). The Miller model, on the contrary, described the proper creep curve, although it did not succeed in predicting the subsequent plastic behavior shown in Fig. 5. It is difficult to determine the complete set of parameters in the Miller model without any contradiction with the experimental results.

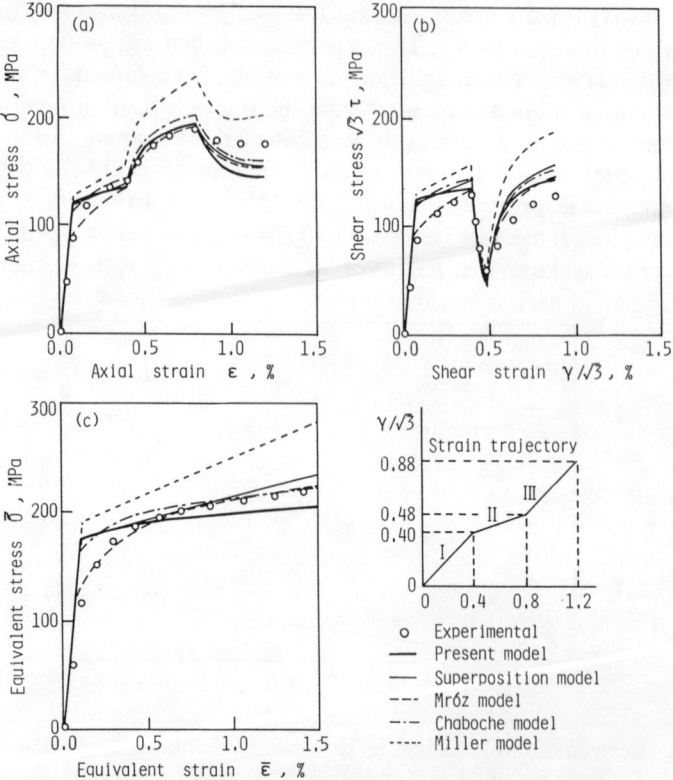

FIG. 7. Stress–strain response of SUS304 steel at 650 °C due to a change in strain trajectory. (a) and (b) indicate the axial and torsional stress–strain relation, respectively, and (c) gives the equivalent stress–strain curve.

## Material Response Due to a Change in Strain Trajectory

Combined proportional strains were applied in the tension–torsion state at the equivalent strain rate of $\bar{\dot{\varepsilon}} = 0\cdot1\%\ \mathrm{s}^{-1}$ for 304 steel as shown in Fig. 7, where a 45° strain was imposed up to an axial strain of 0·4% in stage I, which was followed by stage II with a strain rate ratio of 5 up to $\varepsilon = 0\cdot8\%$. Finally, the direction was again changed to be the same as that in stage I (stage III). Figures 7(a) and 7(b) indicate the axial and torsional stress–strain responses. and Fig. 7(c) gives the equivalent stress–strain curve based on von Mises criterion.

All the models examined could describe a sudden increase in axial stress as well as the steep drop in shear stress at the beginning of stage II, which were reproduced in the experiment. The Mróz model predicted a smooth rounded curve for the stress–strain relation (see Fig. 7(c)), which is appreciated, while the Miller model overestimated the stress response.

## Mechanical Ratcheting Behavior

The accumulation in axial strain due to cyclic shear strain was evaluated under steady and/or varying axial stress levels. Figure 8 shows the accumulated axial strain under an axial stress of $\sigma = 67$ MPa at the shear strain rate of $\dot{\gamma}/\sqrt{3} = 0.002\%$ s$^{-1}$ for SUS304 steel (Fig. 8(a)) and that for Cr–Mo steel under $\sigma = 80$ MPa at $\dot{\gamma}/\sqrt{3} = 0.1\%$ s$^{-1}$ (Fig. 8(b)).

As can be deduced from the figure, all the analytical results for 304 steel give too large a ratcheting strain. The proposed model predicted only one and a half cycles for the accumulation to reach 2%, whereas it took about three and a half cycles in the actual experiment. The results for Cr–Mo steel provide some variation between the simulations, the result from the Mróz model saturating at a certain value of the axial strain while the other models still predicted a large ratcheting strain. These differences may have been due to the variation of back stress in the axial direction during cyclic shear straining.

In order to learn the effect of the axial stress and rate dependence of the material, both the stress level and the shear strain rate were changed during the test. Table 2 summarizes the experimental conditions for $2\frac{1}{4}$Cr–1Mo steel at 600 °C, in which Path A was the case for a drop in axial stress from 60 MPa to 20 MPa, and Path B was the change of strain

TABLE 2
Conditions for the mechanical ratcheting test with varying stress

| Path | Stage I ($\varepsilon \leqslant 1.5\%$) | | Stage II ($\varepsilon \geqslant 1.5\%$) | |
|------|------------------|------------------------------------|-------------------|------------------------------------|
| | $\sigma_I$ (MPa) | $\dot{\gamma}_I/\sqrt{3}$ (% s$^{-1}$) | $\sigma_{II}$ (MPa) | $\dot{\gamma}_{II}/\sqrt{3}$ (% s$^{-1}$) |
| A | 60 | 0.1 | 20 | 0.1 |
| B | 40 | 0.002 | 20 | 0.1 |

rate. Ratcheting behavior can be observed in Fig. 9 as the relation between the axial strain and the number of cycles. The simulated results for Path A show characteristic differences between the models; for

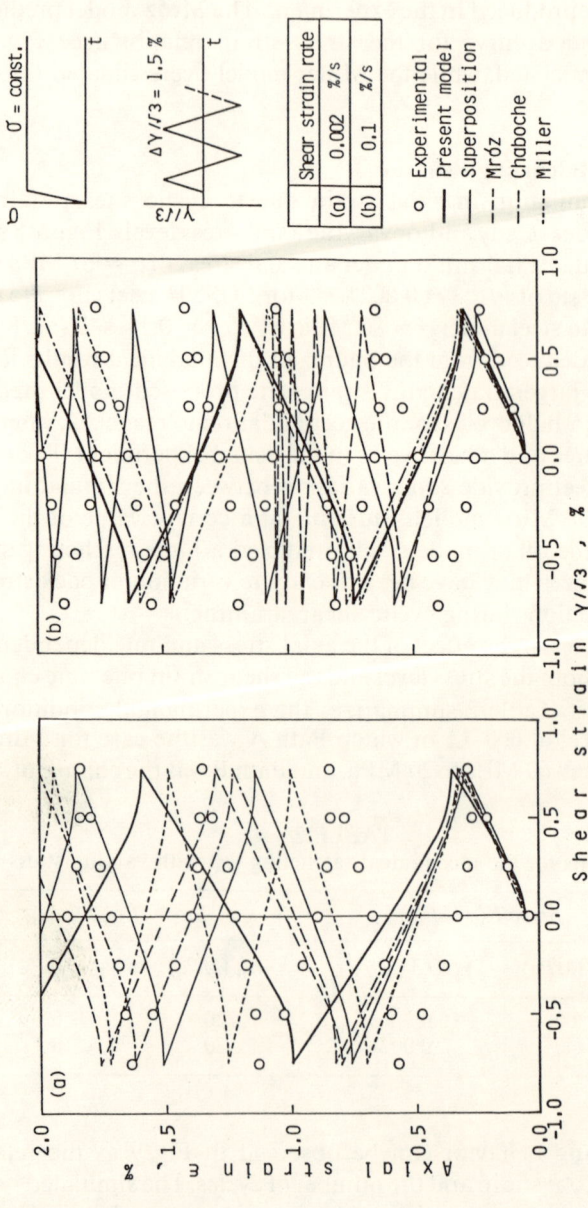

FIG. 8. Relationship between the accumulated axial strain and cyclic shear strain under mechanical ratcheting. (a) is for SUS304 steel at 650°C and (b) is for $2\frac{1}{4}$Cr–1Mo steel at 600°C.

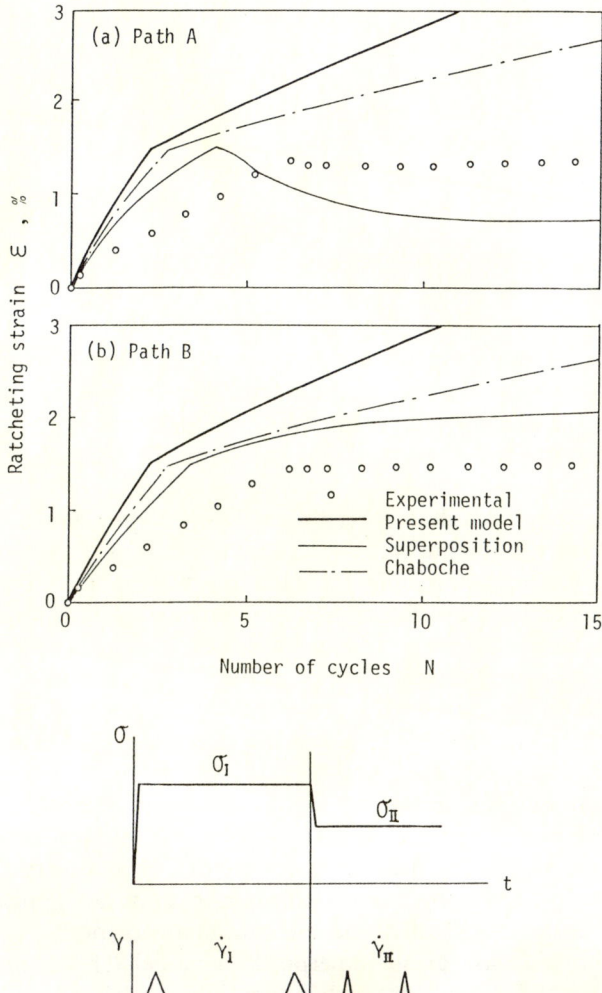

FIG. 9. Relation between the ratcheting strain and the number of cycles for $2\frac{1}{4}$Cr-1Mo steel at 600 °C.

instance, the superposition model may predict some limit in the accumulation level for the axial stress, while the proposed model and the Chaboche model still present a gradual increase in ratcheting strain with cycles despite the drop in axial stress. Both experimental results show that the accumulation of axial strain was completed or even decreased in the early cycles of stage II.

## DISCUSSIONS ON THE CREEP-PLASTICITY INTERACTION BEHAVIOR

The creep–plasticity interaction behavior obtained in the experiments and simulations was particularly governed by the material anisotropy induced by prior deformation, the validity of the normality rule and the variation of internal stresses during mechanical ratcheting.

### Anisotropy Developed during Plastic and Creep Deformations

Both the experimental results illustrated in Figs. 5 and 6 show a hardening phenomenon under prior loading, which is usually regarded as the development of internal stresses such as back stress and drag stress. Such phenomena can be explained for the unified constitutive model as is schematically illustrated in Fig. 2. When a plastic prestrain is imposed on the materials, kinematic hardening as well as isotropic hardening will occur, and therefore the effective stress (excess stress) will decrease under subsequent loading conditions. It follows that a curve such as that shown in Fig. 6 can be predicted by unified constitutive models, although there were some quantitative discrepancies between the simulation and the experiment. Conversely, the effect of preloading on the subsequent deformation may not appear if the structure is composed of a linear combination of plastic and creep elements in the case of the superposition model. This is one of the reasons why the unified approach is superior to the classical method in mixed loading conditions of plasticity and creep.

The development of anisotropy due to preloading is of importance and should be considered as the variation of internal stresses, since all the models examined here were tacitly based on an isotropic condition of the von Mises type. Figure 10 [16] demonstrates several yield loci (with 0·05% inelastic strain) under plastic and creep loading conditions, in which the locus nearly follows the von Mises type (circle) in the virgin state, while the shape changed under progressive deformation. In

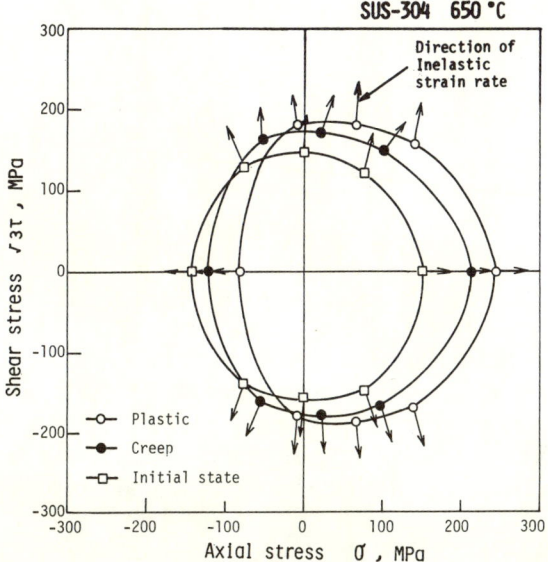

FIG. 10. Subsequent yield loci in tension and creep for SUS304 steel at 650 °C.

particular, the Bauschinger effect (shrinking of the locus in the loading direction) as well as the cross effect (expansion of the circle on the shear side in opposition to the direction) can be clearly seen from the figure, which we have no method for overcoming. It should be noticed that the shape of the yield loci does not depend on the loading condition between plasticity and creep. This problem should preferably be considered in further developments to formulate the constitutive models. It should also be noted that there is some deviation from the normality rule apparent in the figure.

**Stress Response Due to a Change in the Direction of Strain Rate**

Despite the experimentally observed deviation from the normality rule (see Fig. 10), we can explain the reason for the sudden increase and decrease of stress responses shown in Fig. 7 (see Fig. 11). Let us suppose that the stress point $(\sigma_A, \tau_A)$ lies on the yield surface $Y_A$ with $\dot{\varepsilon}_A^i$. When the direction of the strain rate changes from A to B, the stress point should move to $(\sigma_B, \tau_B)$ on $Y_B$ with $\dot{\varepsilon}_B^i$ according to the normality rule. This implies that the component of axial stress increases from $\sigma_A$ to $\sigma_B$, while the shear stress decreases from $\tau_A$ to $\tau_B$. Since the length of the strain path does not have any influence on the corresponding stress

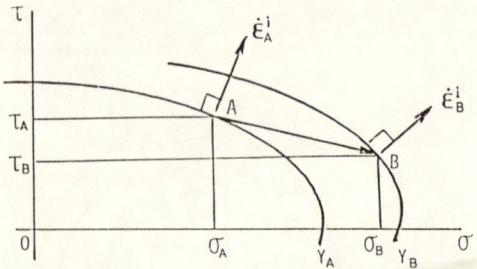

FIG. 11. Yield surfaces with stress point A and B.

variation, the stress must rapidly change to its most suitable level; such a steep change of stress response can be observed in both the experimental and all the simulated results, which show good agreement with the experimental case. This fact suggests that the real material may roughly obey the normality rule for inelastic strain rate at a certain surface in stress space.

### Accumulation of Ratcheting Strain

The feature of mechanical ratcheting behavior is illustrated in Fig. 12(a), where only the kinematic hardening rule was, for convenience, assumed to represent the variation of incremental ratcheting strain, and the normality rule was also employed. When a cyclic shear strain is

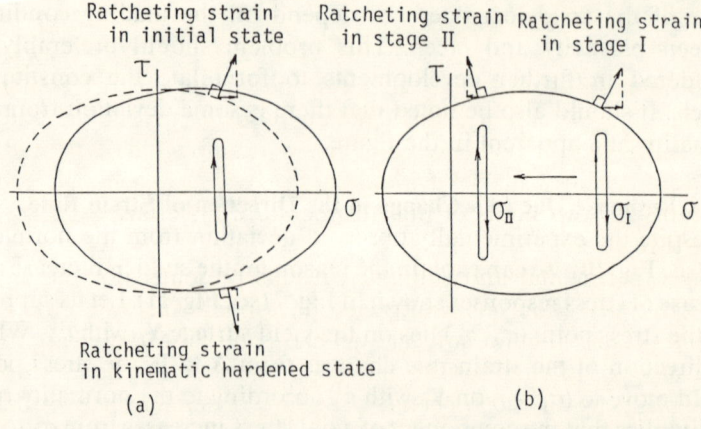

FIG. 12. Illustration of the mechanical ratcheting behavior in stress space. (a) shows that the increment of ratcheting strain becomes smaller with the number of cycles, and (b) indicates that the ratcheting strain decreases despite a positive axial stress $\sigma_{II}$.

imposed on the material under steady axial stress, the axial component of the back stress will increase with the number of cycles, and hence the axial increment of ratcheting strain will gradually decrease with the number of cycles. Moreover, it will also do so if isotropic hardening occurs. This is the mechanical interpretation of the ratcheting phenomenon, although we found an overestimation of ratcheting by the models. One reason for this is the anisotropy induced by a complicated loading regime like ratcheting, since we have already found remarkable anisotropy even in a monotonic prestrained state (see Fig. 10). The other reason may be due to the accuracy in estimating the variation of the back stress, the isotropic variables and stress–strain relation itself. For instance, the back stress employed for the Mróz model in Fig. 8(b) saturated at 80 MPa, which coincides with the axial stress input in the experiment, enabling no more ratcheting strain to be accumulated.

When the axial stress level dropped from $\sigma_I$ to $\sigma_{II}$ in Fig. 12(b), the axial strain increment may have become negative even at a positive axial stress level [17]. In considering the recovery effect of back stress during creep deformation, we could predict that the subsequent ratcheting strain with creep preloading would be larger than that with plastic prestrain. In fact, the experimental results indicating the early cycles in stage II of Figs. 9(a) and 9(b) show such a rate-dependent effect.

## CONCLUDING REMARKS

An inelastic constitutive equation based on the excess stress hypothesis is proposed in this chapter. By using this model and the models developed by other researchers, we examined the capability for describing the inelastic behavior under three series of complex loading programs for two kinds of high temperature materials, SUS304 and $2\frac{1}{4}$Cr–1Mo steel.

The experimental and simulated results presented for the creep-plasticity interaction behavior, material response due to a change in strain trajectory and mechanical ratcheting prove the unified approach to be superior to the classical superposition model for describing the creep–plasticity interaction behavior and the material response under a complicated loading regime.

Phenomena such as material anisotropy, the normality rule and the accumulation of ratcheting strain account for the variation of internal variables.

20     T. INOUE AND S. IMATANI

## ACKNOWLEDGMENTS

The authors wish to express their sincere gratitude to Sumitomo Metal Co. Ltd. and Kawasaki Steel Co. Ltd. for providing the materials used for the experiments. A part of this work was supported by the Ministry of Education, Japan, as a Grant-in-Aid for Cooperative Research A (No. 59350008).

## REFERENCES

[1]  J. M. Corum, W. L. Greenstreet, K. C. Liu, C. E. Pugh and R. W. Swindeman, Rep. No. ORNL-5014, Oak Ridge National Laboratory (1974).
[2]  C. E. Pugh, J. Pressure Vessel Tech., Trans. ASME, 105, 273 (1983).
[3]  P. Perzyna, Adv. Appl. Mech., 11, 313 (1971).
[4]  J. L. Chaboche and G. Rousselier, J. Pressure Vessel Tech., Trans. ASME, 105, 153 (1983).
[5]  K. C. Valanis, Arch. Mech., 23, 517 (1971).
[6]  D. Kujawski and Z. Mróz, Acta Mech., 36, 213 (1980).
[7]  A. Miller, J. Eng. Mater. Technol., Trans. ASME, 98, 97 (1976).
[8]  J. Kratochvil and O. W. Dillon, Jr., J. Appl. Phys., 41, 1470 (1970).
[9]  E. W. Hart, J. Eng. Mater. Technol., Trans. ASME, 98, 193 (1976).
[10] J. H. Gittus, J. Eng. Mater. Technol., Trans. ASME, 98, 52 (1976).
[11] Y. F. Dafalias and E. P. Popov, J. Appl. Mech., Trans. ASME, 43, 645 (1976).
[12] N. Ohno and Y. Kachi, J. Appl. Mech., Trans. ASME, 53, 395 (1986).
[13] N. Ohno, J. Appl. Mech., Trans. ASME, 49, 721 (1982).
[14] E. Tanaka, S. Murakami and M. Ooka, Trans. JSME (A), 52, 2054 (1986).
[15] T. Inoue, S. Imatani and T. Sahashi, Proc. 28th Japan Congr. Mater. Res., 15, Society of Materials Science, Kyoto, Japan (1985).
[16] S. Imatani and T. Inoue, Proc. 2nd Int. Conf. Constitutive Laws for Eng. Mater., 1, 573, Elsevier, Tokyo (1987).
[17] E. Udoguchi, Y. Asada, S. Mitsuhashi and T. Hiroe, Trans. JSME, 42, 1993 (1976).

# Cyclic Stress–Strain Behavior and Life Prediction of $2\frac{1}{4}$Cr–1Mo Steel in a Creep–Fatigue Interaction Regime

TOSHIHIDE IGARI, KATSUYA SETOGUCHI and MASAFUMI YAMAUCHI

*Nagasaki Research and Development Center, Mitsubishi Heavy Industries, Ltd, 1-1 Akunoura-machi, Nagasaki 850-91, Japan*

## ABSTRACT

An experimental study of the creep–plasticity interaction on cyclic stress–strain behavior was performed in a uniaxial stress state on a normalized and tempered $2\frac{1}{4}$Cr–1Mo steel, and the behavior obtained was applied to life prediction in a creep–fatigue interaction regime. First, the correspondence between a cyclic stress–strain curve and a hysteresis curve was examined at a fast strain rate at which creep deformation was negligible. Secondly, the effect of the creep strain history on the cyclic stress–strain curve as well as that of the cyclic plastic strain history on creep were examined to study the creep–plasticity interaction on the cyclic stress–strain behavior. Isotropic softening was found from both the cyclic stress–strain curve and the minimum creep rate, in comparison with the case without creep–plasticity interaction, and the influence of thermal aging on this softening was pointed out. The cyclic stress–strain curve under creep–plasticity interaction was predicted from both the inelastic strain history and the aging duration. Finally, life prediction in a uniaxial creep–fatigue test was made from a hysteresis loop obtained by inelastic analysis, adopting the life fraction rule and the strain range partitioning method for life estimation. As a result, creep–plasticity interaction on the cyclic stress–strain behavior was found to be important for making an accurate life prediction in a creep–fatigue interaction regime.

## INTRODUCTION

In the design of nuclear and fossil-power plant components used at high temperatures, the development of a prediction method for creep–fatigue life based on inelastic analysis is required. In order to make an accurate life prediction, it is necessary to adopt both an inelastic constitutive model describing the creep–plasticity interaction [1] and a life estimation method for describing the creep–fatigue interaction. Many constitutive models under creep–plasticity interaction have been proposed, and a trial to evaluate these models by making a benchmark analysis has been reported [1]. A critical review of the life estimation methods under creep–fatigue interaction has been made [2], and a trial to make a life prediction by combining these constitutive models and life estimation methods has been performed [3].

In applying the life prediction method to real structural components, a representative stress–strain hysteresis loop obtained from inelastic analysis is usually used for life prediction, because inelastic analyses following all cyclic histories are not realistic. In order to obtain a representative hysteresis loop in a creep–fatigue interaction regime, it is necessary to clarify experimentally the cyclic stress–strain curve [4] and the cyclic creep property under the creep–plasticity interaction. The creep–plasticity interaction on a monotonic stress–strain curve and on monotonic creep has been reported [5–9], but this interaction on a cyclic stress–strain curve and on a cyclic creep property for applying to creep–fatigue life prediction has not been studied. It has also remained unanswered whether or not the effect of creep–plasticity interaction on the cyclic stress–strain behavior plays an important role in predicting the creep–fatigue life.

In this chapter an experimental study of the cyclic stress–strain behavior under creep–plasticity interaction was made on a normalized and tempered $2\frac{1}{4}$Cr-1Mo steel (SA387, Gr.22), and the behavior was used to predict the creep–fatigue life. Initially, the correspondence

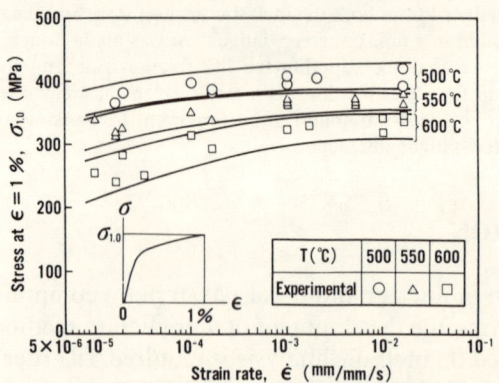

FIG. 1. Strain rate effect on stress at 1% total strain.

between a cyclic stress–strain curve and a hysteresis curve in plasticity is clarified, after determining a strain rate at which creep deformation was negligible. The influence of creep–plasticity interaction on both the cyclic stress–strain curve and the cyclic creep property was subsequently examined, and a prediction method for the cyclic stress–strain curve under creep–plasticity interaction is proposed. Finally, the influence of creep–plasticity interaction on the prediction of creep–fatigue life was studied by adopting the life fraction rule (LFR) and the strain range partitioning (SRP) method for estimating the life. The representative hysteresis loop was determined by an inelastic analysis, using the cyclic stress–strain behavior that was experimentally obtained from the cyclic stress–strain curve and the cyclic creep property.

## CYCLIC PLASTIC PROPERTY AND CYCLIC STRESS–STRAIN CURVE

### Strain Rate Effect

There are two methods used for examining the creep–plasticity interaction [1], that of the unified constitutive model without distinguishing between the plastic and creep strains, and that of the modified superposition model, which distinguishes between these two strains and considers the interaction between them. In this chapter the second method was employed, and it was necessary to confirm the existence of a fast strain rate that could be regarded as 'plastic'. Figure 1 [10] shows the flow stresses at 1% total strain from a monotonic tensile test with constant strain rates at 500, 550 and 600 °C. An increasing tendency of the flow stresses is shown with increasing strain rate, and the increase of the flow stress became saturated at high strain rates. The strain rates $\dot{\varepsilon}_0$ at which the increase of flow stresses was saturated can be regarded as follows:

$$\dot{\varepsilon}_0 = 0.2\% \text{ s}^{-1} \qquad \text{at } 500\,^{\circ}\text{C}$$

$$\dot{\varepsilon}_0 = 0.5\% \text{ s}^{-1} \qquad \text{at } 550\,^{\circ}\text{C}$$

$$\dot{\varepsilon}_0 = 1\% \text{ s}^{-1} \qquad \text{at } 600\,^{\circ}\text{C}$$

Hereafter, the stress–strain behavior at the strain rate $\dot{\varepsilon}_0$ is regarded as plastic, and the expression 'cyclic stress–strain curve' is used for describing the material behavior at the same strain rate.

## Correspondence Between the Cyclic Stress–Strain Curve and the Hysteresis Curve

It was important to clarify whether or not the hysteresis loop in a cycle was reproducible by the cyclic stress–strain curve obtained by the companion specimen test [11].

Figure 2 shows a monotonic stress–strain curve and an ascending portion of the hysteresis curve in a double-logarithmic diagram, where the origin of the coordinates $\sigma$-$\varepsilon_p$ is taken at the compressive tip for the case of the hysteresis loop. As shown in the figure, the stress–strain

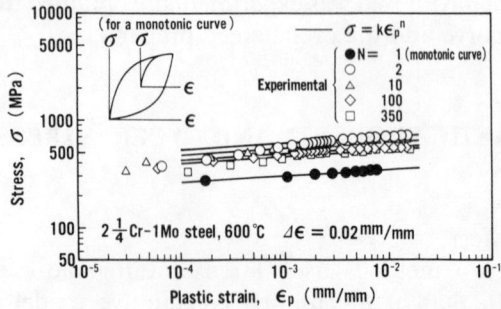

FIG. 2. Example of the relationship between stress and plastic strain in a monotonic stress–strain curve and hysteresis curve.

behavior in the range of plastic strain, $\varepsilon_p$ over 0.2% can be approximated by parallel linear lines with the same slope, so that the $\sigma$-$\varepsilon_p$ relationship in this range can be expressed by Eqn. (1) with the same hardening exponent $n$:

$$\sigma = k \varepsilon_p^{\,n} \tag{1}$$

The cyclic tests with various strain ranges clarified that the value of the hardening exponent $n$ did not depend much on the strain range, and that the change in the exponent $n$ with the number of cycles was small. The change in the coefficient $k$ is plotted in Fig. 3 [12] as a function of the number of cycles $N$, the fatigue damage $N/N_f$ and the accumulated equivalent plastic strain $\bar{\varepsilon}_p$, using a constant value for the exponent $n$. The reason for choosing the first and the second types of abscissa was that the cyclic stress–strain curve is usually defined at a prescribed number of cycles and a prescribed fatigue damage, and the reason for choosing the third type of abscissa was that the effect of cycling is usually considered as a plastic strain history. The strain range did not affect the cyclic softening behavior much for the three types of abscissa. The material property $k$, that was responsible for the shape of the hysteresis loop, together with the hardening exponent, $n$, was found to have the same values independently of the strain range [12] at each stage of number of cycles, fatigue damage and strain history.

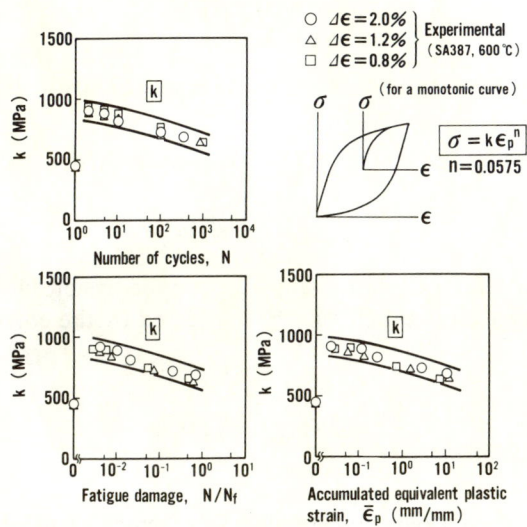

FIG. 3. Example of the change in cyclic hardening strength.

A comparison between the ascending curve of the hysteresis loop scaled by a factor of $\frac{1}{2}$ using the Masing hypothesis [13] and experimentally obtained hysteresis tips is made in Fig. 4 at the 10th cycle, the 100th cycle and the half-life. As indicated in the figure, the scaled hysteresis curves coincide well with the hysteresis tips. The cyclic stress-strain curve from the companion specimen test at the prescribed fatigue damage can be predicted from the hysteresis curve scaled by a factor of $\frac{1}{2}$ at the same cycles or the same fatigue damage. Thus, assuming that the ascending curve of the hysteresis loop is expressed as the following equation in the coordinate system updated at each stress reversal:

$$\sigma = g(\varepsilon_p) = k\varepsilon_p^{\,n} \qquad (2)$$

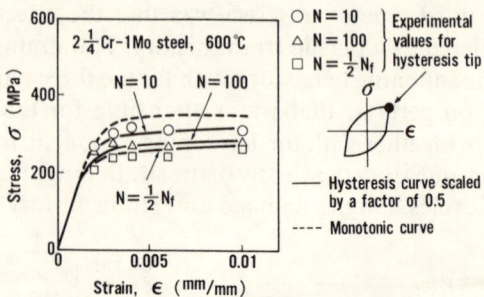

FIG. 4. Comparison between the cyclic stress–strain curve and hysteresis loop.

and also assuming that the cyclic stress–strain curve can be formulated as:

$$\sigma = f(\varepsilon_p) = k' \varepsilon_p^{n'} \tag{3}$$

it was found that the relationship between the hysteresis curve and the cyclic stress–strain curve could be expressed by the equations:

$$n = n' \tag{4}$$

$$2f(\varepsilon_p) = g(2\varepsilon_p) \tag{5}$$

$$k' = 2^{n-1}k \tag{6}$$

In the case of a type 304 stainless steel which exhibits a significant strain range dependence on cyclic hardening behavior, it has been reported [14] that the cyclic stress–strain curve from the companion specimen test did not correspond well with the hysteresis curve. Consequently, the correspondence between the cyclic stress–strain curve and the hysteresis curve just discussed seems to be due to the fact that the cyclic softening behavior of the test material adopted here did not depend on the strain range.

## CREEP–PLASTICITY INTERACTION ON THE CYCLIC STRESS–STRAIN BEHAVIOR

### Influence on the Cyclic Stress–Strain Curve

Initially, the influence of creep prestrain on subsequent cyclic plastic behavior was examined. Figure 5 indicates the time histories of peak stresses from plastic cycling tests at 550°C after creep tests, in comparison with a plastic cycling test without creep prestrain [15]. The conditions for the pre-creep test at 550°C were as follows:

(1) monotonic creep for 8 h at a stress of 176 MPa ($\varepsilon_c$ = 0·236%);
(2) monotonic creep for 8 h at a stress of 216 MPa ($\varepsilon_c$ = 0·425%);
(3) monotonic creep for 8 h at a stress of 245 MPa ($\varepsilon_c$ = 0·886%);
(4) reversed creep for 50 cycles with hold time of 53 min at a stress of ±176 MPa ($\bar{\varepsilon}_c$ = 80·31%, where $\bar{\varepsilon}_c$ is the accumulated creep strain).

Peak stresses after the creep tests were found to be smaller in the initial cycles than those without creep prestrain, showing little difference between the tension and compression sides. The reduction of peak stresses after test (4), in which the material was subjected to the highest creep prestrain with longer thermal exposure time, was the largest.

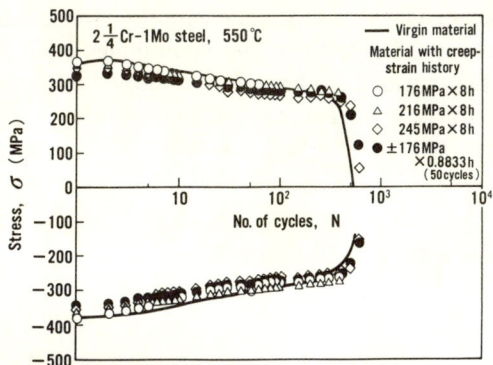

FIG. 5. Influence of pre-creep strain on the subsequent cyclic stress amplitude.

In the creep-fatigue conditions for a real component, in addition to the sequence effect of plastic and creep strains just shown, the case when plastic and creep strains are simultaneously included in a strain cycle is important, as in the fundamental tests of the SRP method by Manson [16]. Here, the relationship between the plastic property in such a strain cycle and the inelastic strain history was examined, assuming that the stress-strain behavior of rapid straining adopted in the SRP method was fully plastic. Figure 6 indicates an example of a hysteresis loop with

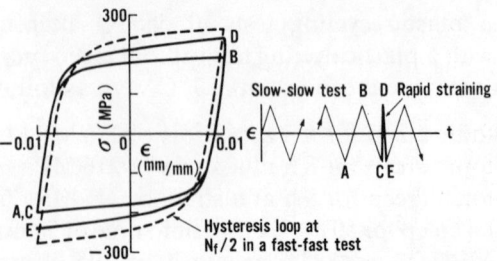

FIG. 6. Example of a hysteresis loop obtained by rapid straining in a slow–slow test.

rapid straining at around $N_f/2$ cycles, which was obtained from a slow–slow test ($\dot{\varepsilon} = 0.01\%$ s$^{-1}$) at 600 °C with rapid straining of the strain rate ($\dot{\varepsilon}_0 = 1\%$ s$^{-1}$) inserted at every 50 cycles. A hysteresis loop at $N_f/2$ cycles in a fast–fast test with the strain rate of $\dot{\varepsilon}_0$ is also shown in the figure by the broken line, and it follows that the hysteresis loop obtained by rapid straining shows an isotropically softened property compared with that obtained in the fast–fast test at the state with the same damage. Figure 7 expresses the relationship between the accumulated equivalent inelastic strain $\bar{\varepsilon}_{in}$ and the material coefficient $k$, obtained by approximating a hysteresis loop for rapid straining in the slow–slow tests as Eqn. (1), and comparing them with a fast–fast test in which $\bar{\varepsilon}_p$ was regarded as $\bar{\varepsilon}_{in}$.

$$\bar{\varepsilon}_{in} = \bar{\varepsilon}_p + \bar{\varepsilon}_c \tag{7}$$

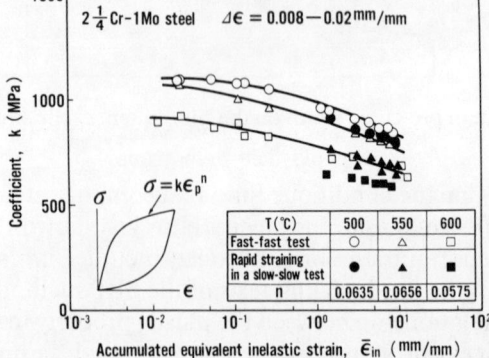

| T (°C) | 500 | 550 | 600 |
|---|---|---|---|
| Fast-fast test | O | △ | □ |
| Rapid straining in a slow-slow test | ● | ▲ | ■ |
| n | 0.0635 | 0.0656 | 0.0575 |

FIG. 7. Change in cyclic hardening strength with accumulated equivalent inelastic strain $\bar{\varepsilon}_{in}$ under creep–plasticity interaction.

The figure shows that the $k$ values at the same $\bar{\varepsilon}_{in}$ in the rapid straining and fast–fast tests differed from each other, the $k$ values obtained from rapid straining being smaller than those from the fast–fast test. This fact indicates that the $k$ values in the case with combined plastic and creep strains could not be predicted by the combination of the $k$ values in the fast–fast tests and the inelastic strain history parameter $\bar{\varepsilon}_{in}$.

We adopted the Manson–Succop-type parameter approach in order to take account of the creep–plasticity interaction as a combination of the inelastic strain history and time history for thermal aging, considering that the material was subjected to both histories in the slow–slow test. The following parameter $P_1$ was originally proposed to treat creep damage by Goldhoff and Woodford [17]:

$$P_1 = 0.01T + \log(\varepsilon t) \tag{8}$$

The data in Fig. 7 were plotted in Fig. 8, taking the abscissa as a logarithmic scale of parameter $P_2$ for each temperature

$$P_2 = \bar{\varepsilon}_{in}t \tag{9}$$

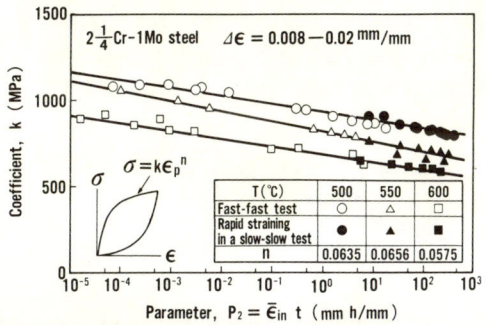

FIG. 8. Change in cyclic hardening strength with the parameter $\bar{\varepsilon}_{in}t$ under creep–plasticity interaction.

in which the cyclic softening of the test material under creep–plasticity interaction was controlled by the time history for thermal aging in addition to the inelastic strain history. It was found from the figure that the data from both the fast–fast test and rapid straining could be expressed by one master curve, and that there was a possibility for predicting the plastic property of the material which experienced arbitrary strain and time histories under the creep–plasticity interaction.

## Influence on Creep Behavior

Ohtani *et al.* [18] have reported that the creep property changed with advancing cycles in creep–fatigue tests and was different from the monotonic creep property, but a quantitative study of this phenomenon has not been made. Here, as a first step towards examining the change of creep property with progressive cycles in creep–fatigue tests, a comparison between the results of a creep test after plastic cycling and those of a monotonic creep test was made. The results of creep tests after plastic cycling of 10 and 200 cycles (near $N_f/2$) with a strain range of 2% at 550 °C were compared with those of the monotonic creep test without pre-plastic cycling (see Fig. 9 [15]). Plastic straining for 10 cycles was found to have little effect on the subsequent creep behavior, but that for 200 cycles represented a softening effect on the subsequent creep by accelerating the creep strain rate.

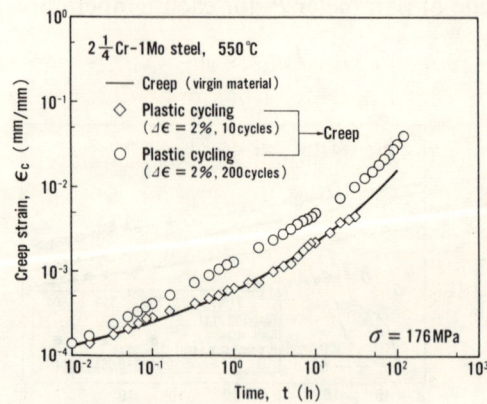

FIG. 9. Influence of cyclic pre-plastic strain on subsequent creep.

In order to consider the influence of the time history of thermal aging already indicated, creep tests under the stress level of 176 MPa at 550 °C after thermal aging for 200, 500, 1000, 2000 and 5000 h without stress at 550 °C were carried out. The results of the tests are summarized in Fig. 10, in comparison with those of the monotonic creep test at the same stress. The creep curves after thermal aging, except for the case of 200 h, show a softening tendency compared with the curve without pre-thermal aging, and the creep strain rate was accelerated in the case of longer thermal aging.

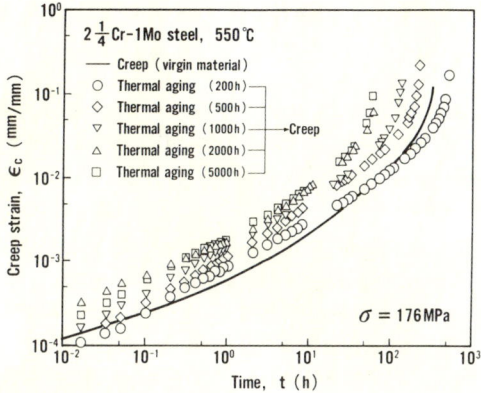

FIG. 10. Influence of thermal aging on subsequent creep.

The minimum creep rates obtained from the creep tests after plastic cycling for 200 cycles (about $N_f/2$) with a strain range of 2% at 500, 550 and 600 °C are plotted against the Larson–Miller parameter (LMP) in comparison with those obtained from monotonic creep tests in Fig. 11. The minimum creep rates after cyclic plastic strain histories generally became faster than those in monotonic creep, and were much different at higher stresses while little different at low stresses of around 120 MPa.

The minimum creep rates obtained from the creep tests after thermal aging are also shown in this figure. These minimum creep rates were also faster than those in the monotonic creep tests, and the degree of acceleration of these rates was about the same as that from the creep tests after pre-plastic cycling.

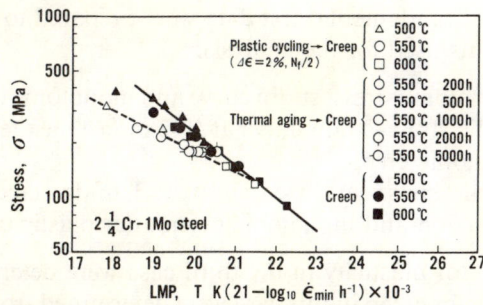

FIG. 11. Minimum creep rates under creep–plasticity interaction.

## CREEP–FATIGUE LIFE PREDICTION BASED ON INELASTIC ANALYSIS

A creep–fatigue life prediction based on inelastic analysis using the experimentally obtained cyclic stress–strain behavior was made for a fully reversed uniaxial strain hold test at 600 °C. The test conditions and failure lives were as follows:

(1) $\Delta\varepsilon$ = 2%, hold time $t_h$ in tension = 13·4 min, $N_f$ = 410;
(2) $\Delta\varepsilon$ = 1%, $t_h$ in tension = 32·4 min, $N_f$ = 820;
(3) $\Delta\varepsilon$ = 1%, $t_h$ in tension = 5·4 min, $N_f$ = 940;
(4) $\Delta\varepsilon$ = 1%, $t_h$ in tension = 0·9 min, $N_f$ = 1070;
(5) $\Delta\varepsilon$ = 2%, $t_h$ in compression = 13·4 min, $N_f$ = 472;
(6) $\Delta\varepsilon$ = 1%, $t_h$ in compression = 13·5 min, $N_f$ = 1360.

Here, the strain rate except for the strain hold period was kept constant at 1% s$^{-1}$.

The simple superposition model with independent plastic and creep strains, and neglecting the creep–plasticity interaction, was employed for the analysis. A combination of the kinematic hardening model without cyclic softening and the Masing model [12, 13] following Eqn. (5) were used for plasticity, and the creep equation in Eqn. (10) was adopted by modifying the strain hardening rule according to ORNL:

$$\varepsilon_c = a\sigma^b t^c + d\sigma^e t \qquad (10)$$

where $a$–$e$ are constants. The strain rate effect could be determined by performing an elasto-plastic analysis with a subsequent creep analysis in a prescribed time step. The effectiveness of this scheme to express the strain rate effect has been confirmed by Inoue et al. [1]. An analysis for two cycles was carried out to obtain the representative hysteresis loop.

Three kinds of fundamental test data were prepared to identify the material constants used for the analysis:

Case 1, a monotonic stress–strain curve and monotonic creep data;
Case 2, a cyclic stress–strain curve at $N_f/2$ in fast–fast test as well as monotonic creep;
Case 3, a cyclic stress–strain curve at $N_f/2$ under creep–plasticity interaction and monotonic creep after plastic cycling.

The parameters for plasticity in the third case were determined from Fig. 8, and those for creep in this case were determined from Fig. 11 by approximating the secondary creep as a dotted line (for $\sigma \geqslant 120$ MPa).

TABLE 1
Material constants used in the inelastic analysis

| | Plasticity | | Creep | | | | |
|---|---|---|---|---|---|---|---|
| | $k$ (MPa) | $n$ | $a$ | $b$ | $c$ | $d$ | $e$ |
| Case 1 | 453·1 | | | | | $1·57 \times 10^{-17}$ | 6·33 |
| Case 2 | 367·7 | 0·057 5 | $6·55 \times 10^{-13}$ | 4·36 | 0·502 | $1·57 \times 10^{-17}$ | 6·33 |
| Case 3 | 357·1 − 23·3 $\log_{10}(\bar{\varepsilon}_{in} t)$ | | | | | $(4·32 \times 10^{-25})^*$ | $(9·97)^*$ |

Units: $\varepsilon$, mm mm$^{-1}$; $t$, h.
*Values for $\sigma \geqslant 120$ MPa.

The values of the material constants for each of the three cases are listed in Table 1.

Two types of creep–fatigue life estimation methods, the LFR proposed by Robinson [19] and Taira [20], and the SRP method proposed by Manson [16], were employed. The LFR method, employed in *ASME Boiler and Pressure Vessel Code Case N-47* [21], assumes that material failure occurs when the sum of the fatigue damage $D_f$ as a cycle fraction and the creep damage $D_c$ as a time fraction reaches a characteristic value $D$, ($D = 1$ is employed here) as follows:

$$D_f + D_c = D \tag{11}$$

The SRP method distinguishes the plastic strain related to transgranular damage from creep strain related to intergranular damage, and the inelastic strain range $\Delta\varepsilon_{in}$ is assumed to consist of four components ($\Delta\varepsilon_{pp}, \Delta\varepsilon_{cp}, \Delta\varepsilon_{pc}$ and $\Delta\varepsilon_{cc}$), which are respectively connected with inherent material lives ($N_{pp}, N_{cp}, N_{pc}$ and $N_{cc}$) by the following Manson–Coffin equations:

$$\begin{aligned} \Delta\varepsilon_{pp}N_{pp}{}^{\alpha} = C_1 \qquad \Delta\varepsilon_{cp}N_{cp}{}^{\beta} = C_2 \\ \Delta\varepsilon_{pc}N_{pc}{}^{\gamma} = C_3 \qquad \Delta\varepsilon_{cc}N_{cc}{}^{\delta} = C_4 \end{aligned} \tag{12}$$

The creep–fatigue life for cycling of an arbitrary inelastic strain range was estimated by the 'interaction damage rule [16]' that can be expressed by:

$$1/N_f = \Sigma(F_{ij}/N_{ij}) \tag{13}$$

where

$$\begin{aligned} F_{ij} = \Delta\varepsilon_{ij}/\Delta\varepsilon_{in} \\ \Delta\varepsilon_{in} = \Sigma\Delta\varepsilon_{ij} \qquad i,j = p,c \end{aligned} \tag{14}$$

Examples of the hysteresis loop obtained from inelastic analyses using the three types of material parameters identified by the three methods are shown in Fig. 12, with an experimentally obtained hysteresis loop for the test condition (1) at $N_f/2$. The hysteresis loops based on the methods of Cases 1 and 2 predicted a higher peak stress than the experiment. On the contrary, the hysteresis loop from Case 3 was found to be in a good agreement with the experiment. Residual stresses after relaxation at the strain hold time simulated by the analysis

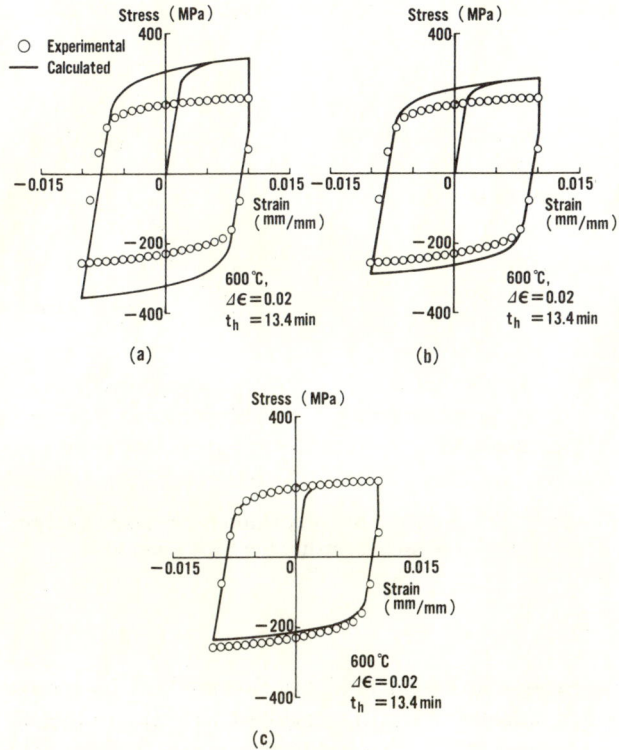

FIG. 12. Examples of analytically obtained hysteresis loops for a creep–fatigue test: (a) Case 1; (b) Case 2; (c) Case 3.

decreased in the order of Cases 1, 2 and 3, but the residual stress in Case 3 was still predicted to have a slightly higher value.

The results of a creep–fatigue life prediction by inelastic analysis for these same test conditions, adopting LFR and the SRP method for life estimation, are presented in Fig. 13. With LFR, the scatter in predicted life based on three types of material parameters was generally large, and a shorter life was predicted than that observed. The predicted life based on Case 3 was around a factor of 3 on life, this being the best prediction. On the contrary, the SRP method produced a smaller scatter of life prediction for the three cases, and all data lay within a scatter band of factors of 1·5. For the test conditions employed here, the LFR was found to be more sensitive to the analytical results in predicting creep–fatigue lives than the SRP method.

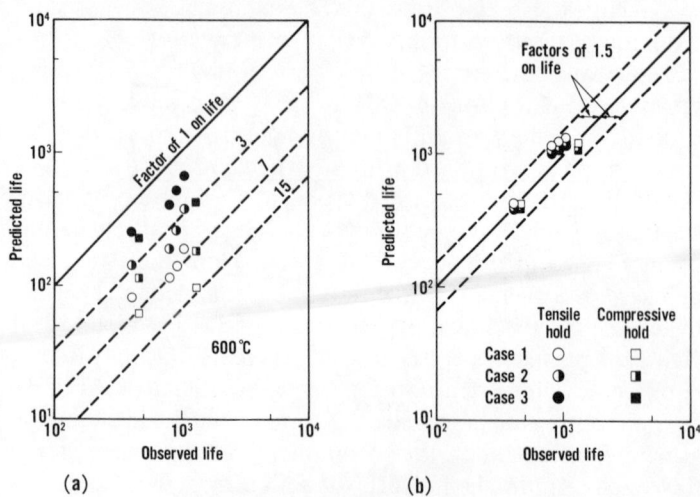

FIG. 13. Results of creep–fatigue life prediction by inelastic analysis (a) using the LFR and (b) using the SRP method.

## DISCUSSION

### The Cyclic Stress–Strain Curve Under Creep–Plasticity Interaction

The fact that creep prestrain caused a hardening in the subsequent monotonic stress–strain curves has been reported by Pugh [5] for a type 304 stainless steel, and by Ohashi et al. [6] for a type 316 stainless steel. On the contrary, Inoue et al. [9] pointed out a softening in the monotonic stress–strain curve for a normalized and tempered $2\frac{1}{4}$Cr–1Mo steel. Concerning the effect of creep prestrain on the subsequent cyclic plastic behavior, a reduction of cyclic hardening in the cyclic plastic behavior of an annealed $2\frac{1}{4}$Cr–1Mo steel in comparison with the case without creep prestrain has been investigated by Jaske et al. [22].

The effect of creep prestrain on subsequent cyclic plastic behavior for the test material adopted here was represented by the softening characteristics of subsequent peak stresses. These softening characteristics, which seemed to be isotropic as indicated by Inoue et al. [23], became greater as the creep prestrain grew larger.

The influence of the combined cyclic plastic and creep strains on subsequent plasticity, which is of importance in obtaining the cyclic stress–strain curve in the creep–plasticity interaction regime, had not been clarified. That is why a trial to predict the cyclic stress–strain curve in such a regime was performed in this chapter by choosing the slow-slow test as a representative strain cycle in the creep-fatigue interaction regime, and by regarding the hysteresis loop for rapid straining as a plastic hysteresis loop. As shown in Fig. 7, for a prescribed inelastic strain history $\bar{\varepsilon}_{in}$, the hysteresis loop for rapid straining was found to have an isotropically softened property when compared with that from the fast-fast test. It follows that the plastic hysteresis loop cannot be predicted by a combination of the cyclic softening property in plasticity shown in Fig. 3 and the accumulated equivalent inelastic strain $\bar{\varepsilon}_{in}$, which regards the plastic and creep strains as equivalent. The trial to use $\bar{\varepsilon}_{in}t$ instead of $\bar{\varepsilon}_{in}$, as shown in Fig. 8, is suggested to be successful in predicting the plastic hysteresis loop under creep-plasticity interaction, based on the cyclic softening property of plasticity (see Fig. 3). The reason for adopting the parameter $\bar{\varepsilon}_{in}t$ comes from the idea that thermal aging, whose influence on the mechanical behavior of material was emphasized by Krempl [24], affects the softening of the material as well as the inelastic strain history. Since the strain range dependence of cyclic softening behavior seems to be small as seen in Fig. 8, the cyclic stress-strain curve under creep-plasticity interaction can be successfully predicted from the figure by using Eqn. (5).

### Creep Behavior Under Creep-Plasticity Interaction

The plastic prestrain effect has been reported by Ohashi et al. [7] to cause hardening in the subsequent monotonic creep of a type 316 stainless steel, and by Inoue et al. [9] of a normalized and tempered $2\frac{1}{4}$Cr-1Mo steel. As for the effect of the cyclic-plastic prestrain on subsequent monotonic creep, hardening in the subsequent creep has been pointed out by Jaske et al. [22] and Inoue et al. [23], who respectively used an annealed $2\frac{1}{4}$Cr-1Mo steel and type 304 stainless steel.

The effect of cyclic-plastic prestrain on subsequent monotonic creep of the present test material was characterized as softening. In this chapter minimum creep rates in subsequent monotonic creep were obtained at various stresses and temperatures in order to examine quantitatively the isotropic softening. An acceleration of the minimum creep rates by cyclic-plastic prestrain is shown in Fig. 11, and the degree of acceleration was found to depend on stress levels. The effect of thermal aging, whose importance was emphasized by Krempl [24], was found to be an acceleration of the minimum creep rate in subsequent creep for the test material. Just as was the case for the cyclic stress-strain curve under creep–plasticity interaction, the cyclic creep property in a creep–fatigue strain cycle that includes both plastic and creep strains simultaneously is presumed to depend on both the inelastic strain history and the thermal aging history. A quantitative prediction of the cyclic creep property should be made in the future.

## Creep–Fatigue Life Prediction Based on Inelastic Analysis

The accuracy for predicting representative hysteresis loops by inelastic analysis using the cyclic stress–strain behavior depends both on the prediction of the degree of cyclic softening in the stress–strain behavior and on the constitutive model employed. With the inelastic analysis adopted here, the difference between the hysteresis loops simulated by three types of constitutive models (i.e. the simple super-position, modified superposition and unified models) has been reported [1] to be small, and consequently the difference between the analytical and experimental hysteresis loops shown in Fig. 12 comes mainly from the accuracy in predicting the degree of isotropic softening just mentioned.

The monotonic stress–strain curve and monotonic creep in Case 1, which are used for the fundamental data in *ASME Boiler and Pressure Vessel Code Case N-47* [21], predict the hysteresis loop with the largest stress range because the softening was neglected. The combination of the cyclic stress–strain curve from the fast–fast test and monotonic creep in Case 2 gave a smaller stress range than that in Case 1, but predicted a larger stress range than the experimental results, since the effect of softening by thermal aging was not considered. The method of Case 3, in which the effect of thermal aging as well as the inelastic strain history were taken into account, gave an adequate hysteresis loop corresponding to the experimental data, but nevertheless showed a slightly higher residual stress in relaxation than the experimental data. A cyclic stress–strain curve under the combined history of inelastic strain and thermal aging can be predicted well by the method shown in this chapter. In order to make a better prediction of the relaxation behavior, consideration should be given to the effects of thermal aging in addition to those of the inelastic strain history.

The life prediction accuracy using the three kinds of analytical hysteresis loops is represented by a large scatter in the case of the LFR, but little scatter in the case of the SRP method. This tendency comes from the following fact as pointed out by the authors [10, 25]; all tests in this chapter were made under strain-controlled conditions, where the difference in the analytical hysteresis loops appeared mainly in the stresses. Consequently, the scatter in the life predicted by the LFR, in which creep damage depends strongly on stress, was large compared with that predicted by SRP, which estimates material lives based on strain. In general, in predicting the life of real structures by inelastic analysis based on the three kinds of material parameters, both ranges of the stress and strain in the analytical hysteresis loops can be expected to be different in each of the three cases, and as a result there can be a large scatter of life prediction even by the SRP method.


## CONCLUDING REMARKS

The cyclic stress-strain behavior under creep-plasticity interaction was investigated to enable life prediction by an inelastic analysis in the creep-fatigue interaction regime, for a normalized and tempered $2\frac{1}{4}$Cr-1Mo steel.

The cyclic stress-strain curve with a fast strain rate, at which creep deformation was negligible, could be simply predicted by a hysteresis curve scaled by a factor of $\frac{1}{2}$. The cyclic stress-strain curve under creep-plasticity interaction was found to have a softer property than this case. A method to predict the cyclic stress-strain curve under creep-plasticity interaction was proposed, which considered the contribution of thermal aging, as well as the inelastic strain history to softening. The creep behavior under creep-plasticity interaction was revealed to have a softened property compared with that under monotonic creep, and both the inelastic strain history and the thermal aging contributed to the softening. An accurate prediction of the cyclic stress-strain behavior under creep-plasticity interaction is important for making an accurate life prediction in the creep-fatigue interaction regime.


## ACKNOWLEDGMENTS

The authors would like to express their gratitude to Messrs. S. Kitade, K. Nakashima and T. Matsubara, members of the Nagasaki Research and Development Center of Mitsubishi Heavy Industries, Ltd., for their cooperation in preparing this paper.

# REFERENCES

[1]   T. Inoue, T. Igari, F. Yoshida, A. Suzuki and S. Murakami, Inelastic behavior of $2\frac{1}{4}$Cr-1Mo steel under plasticity–creep interaction condition, *Nuc. Eng. Des.,* **90**, 287 (1985).

[2]   S. S. Manson, A critical review of predictive methods for treatment of time-dependent metal fatigue at high temperatures, *Pressure Vessels and Piping Design Technology — 1982, A Decade of Progress,* S. Y. Zamrik and D. Dietrich, Eds., ASME, New York, 203 (1982).

[3]   T. Inoue, T. Igari, M. Okazaki, M. Sakane and K. Tokimasa, Fatigue-creep life prediction of $2\frac{1}{4}$Cr-1Mo steel by inelastic analysis — results of joint work, to be presented at *SMiRT-9, Lausanne, Switzerland, August 17–21, 1987.*

[4]   JoDean Morrow, Cyclic plastic strain energy and fatigue of metals, *ASTM STP 378,* ASTM, Philadelphia, PA, 45 (1965).

[5]   C. E. Pugh, Progress in developing constitutive equations for inelastic design analysis, *J. Press. Vessel Technol.,* **105**, 273 (1983).

[6]   Y. Ohashi, M. Kawai and H. Shimizu, Effects of prior creep on subsequent plasticity of type 316 stainless steel at elevated temperature, *J. Eng. Mater. Technol.,* **105** (4), 257 (1983).

[7]   Y. Ohashi, M. Kawai and T. Momose, Effects of prior plasticity on subsequent creep of type 316 stainless steel at elevated temperature, *J. Eng. Mater. Technol.,* **108** (1), 68 (1986).

[8]   Y. Niitsu and K. Ikegami, Plastic and creep deformation of stainless steel SUS 304 under combined stress states at 600 °C, *Trans. JSME,* **50** (453A), 996 (1984) (in Japanese).

[9]   T. Sahashi, S. Imatani and T. Inoue, A constitutive model accounting for the plasticity–creep interaction condition, and the description of inelastic behavior of $2\frac{1}{4}$Cr-1Mo steel, *Trans. JSME,* **52** (476A), 1126 (1986) (in Japanese).

[10]  K. Setoguchi, T. Igari and M. Yamauchi, Creep-fatigue life prediction of $2\frac{1}{4}$Cr-1Mo steel based on inelastic analysis, *J. Soc. of Mater. Sci., Jpn.,* **33** (370), 862 (1984) (in Japanese).

[11]  R. W. Landgraf, JoDean Morrow and T. Endo, Determination of the cyclic stress–strain curve, *J. Mater.,* **4** (1), 176 (1969).

[12]  T. Igari, K. Setoguchi and M. Yamauchi, Study on elastic–plastic deformation analysis using a cyclic stress–strain curve, *J. Soc. of Mater. Sci., Jpn.,* **32** (357), 610 (1983) (in Japanese).

[13]  G. Masing, Eigenspannungen und Verfestigung beim Messing, *Proc. 2nd Int. Cong. of Applied Mechanics, Zurich,* 332 (1926).

[14]  M. Ohnami, M. Sakane and S. Nishino, Cyclic behavior for a type 304 stainless steel in biaxial stress states at elevated temperature, *Int. J. Plasticity,* to be published.

[15]  T. Igari, K. Setoguchi and Y. Wakamatsu, Plasticity–creep interaction and relaxation behavior of $2\frac{1}{4}$Cr-1Mo steel, *Proc. 19th Symp. on High Temperature Strength,* The Society of Materials Science, Japan (1981) (in Japanese).

[16] S. S. Manson, The challenge to unify treatment of high temperature fatigue — a partisan proposal based on strain range partitioning, *ASTM STP 520*, ASTM, Philadelphia, PA, 744 (1972).

[17] R. M. Goldhoff and D. A. Woodford, The evaluation of creep damage in a Cr-Mo-V steel, *ASTM STP 515*, ASTM, Philadelphia, PA, 89 (1972).

[18] R. Ohtani, Ed., Study on high-temperature low-cycle fatigue characteristics of 18Cr-8Ni stainless steel by the strainrange partitioning method, *Report of High-Temperature Strength Research Committee of Iron and Steel Institute of Japan, Tokyo*, (1981) (in Japanese).

[19] E. L. Robinson, Effect of temperature variation on the long-time rupture strength of steels, *Trans. ASME*, **74**, 777 (1952).

[20] S. Taira, Lifetime of structures subjected to varying load and temperature, *Creep in Structures, N. J. Hoff, Ed., Springer, Berlin, 119 (1962)*.

[21] *ASME Boiler and Pressure Vessel Code Case N-47-23*, ASME, New York, 89 (1986).

[22] C. E. Jaske, B. N. Leis and C. E. Pugh, Monotonic and cyclic stress-strain response of annealed $2\frac{1}{4}$Cr-1Mo steel, *Proc. Symp. on Structural Materials for Service at Elevated Temperature in Nuclear Power Generation, Annual Meeting of ASME, USA, November 30-December 5*, 191 (1975).

[23] S. Imatani, H. Haga and T. Inoue, A study on inelastic behavior of SUS 304 under multiaxial stress state — 2nd report, *Trans. JSME*, **52** (437A), 58 (1986) (in Japanese).

[24] E. Krempl, Viscoplasticity based on total strain — the modelling of creep with special considerations of initial strain and aging, *J. Eng. Mater. Technol.*, **101**, 380 (1979).

[25] M. Yamauchi, T. Igari, K. Setoguchi and H. Yamanouchi, Comparison of creep-fatigue life prediction by LFR and SRP methods, *Proc. ASTM Symp. on Low Cycle Fatigue — Direction for the Future, Lake George, New York, September 30-October 4 1985, in ASTM STP 942*, to be published.

# Continuum Theory of Material Damage at High Temperature

SUMIO MURAKAMI

*Department of Mechanical Engineering, Nagoya University, Furo-cho, Chikusa-ku, Nagoya 464, Japan*

and

NOBUTADA OHNO

*Department of Energy Engineering, Toyohashi University of Technology, Tempaku-cho, Toyohashi 440, Japan*

## ABSTRACT

The applicability of continuum damage mechanics to the damage–fracture analysis and life prediction of high temperature machine components was examined. After reviewing the notion and procedure of continuum damage mechanics, its usefulness and potential were compared with those of conventional approaches. The mechanical description of material damage in terms of a tensorial variable, and the derivation of constitutive and evolution equations for damaged materials are discussed in detail, with the anisotropic features of material damage being especially emphasized. To illustrate the scope of the continuum damage approach, the creep rupture process in thin-walled tubes, creep crack growth under complex loading, and the effects of prior plastic damage on the subsequent process of creep rupture were each analyzed.

## INTRODUCTION

The integrity and proper life prediction of high temperature components necessitate a detailed analysis and simulation of damage in the fracture process of material elements under a complicated history of loading.

In conventional procedures for predicting the life of such components, linear and nonlinear fracture mechanics and phenomenological fracture laws expressed in terms of the stress and strain or of their cumulative effects have played the most important roles. Fracture mechanics, above all, has succeeded in providing a powerful means to describe crack growth rates in terms of global fracture parameters. However, while fracture mechanics postulates discrete macroscopic cracks in intact non-dissipative materials, the damaged and inelastic zones of significant size usually precede the crack tips. Besides these limitations in its applicability to ductile fracture and creep crack growth, fracture mechanics cannot be applied to the modelling and analysis of various important problems of damage and fracture, e.g. the process of crack initiation, the effects of local material damage on crack growth, the degradation of rigidity, ductility and toughness as a result of material damage, and the coupling between material damage due to mechanical effects and physicochemical damage by corrosion and irradiation. These limitations also apply to other conventional laws of fracture. This fact implies the necessity for a systematic and versatile approach that is effective to the general process of material damage ranging from microcrack initiation to macroscopic crack growth, or from mechanical damage to physicochemical damage.

Recently, a new systematic discipline has been developed to describe the general process of material damage within the framework of continuum mechanics and on the bases of physics and micromechanics: this discipline is called *continuum damage mechanics* (CDM) (or damage mechanics) [1–11].

The present chapter is concerned with the notion and applicability of CDM as a systematic approach to life prediction and to damage and fracture analysis. Since the geometry, orientation and arrangement of voids and cracks depend on the direction of the applied stress and or strain, the material damage characterized by these cavities generally exhibits significant anisotropy. Thus, after a brief review of the principles and procedures of CDM, a mechanical description of anisotropic damage is discussed by incorporating a fictitious undamaged configuration and a fictitious deformation gradient. The formulation of the evolution and constitutive equations for anisotropic damage will be discussed in some detail. The applicability of CDM will be demonstrated by applying the resulting equations to three typical problems of damage and fracture: the process of creep damage and fracture of thin-walled tubes under complex and non-proportional loadings, the finite element analysis of creep crack growth in copper plates under biaxial non-proportional loadings, and the effect of prior plastic damage on the subsequent process of creep damage.

## LIFE PREDICTION AND DAMAGE ANALYSIS BY MEANS OF CONTINUUM DAMAGE MECHANICS

Let us consider a body B which contains distributed microcavities as shown in Fig. 1 [10], and consider an arbitrary point **x** in the body. We assume that a volume element $V$ exists around **x** which is large enough to include a sufficient number of cavities, but is sufficiently small, at the same time, so that the states of stress, strain and cavity distribution are sufficiently uniform. The notion of CDM hypothesizes that the effects of these microcavities at an arbitrary point **x** can be described by appropriate mechanical variables $D(\mathbf{x})$ or $\omega(\mathbf{x})$, and we call them the *damage fields*. The existence of such damage fields can be admitted according to the same notion as that of a stress, strain or temperature

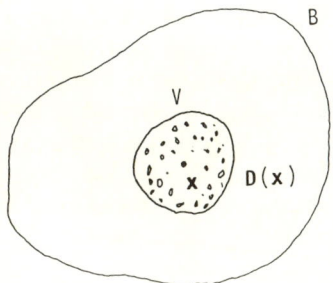

FIG. 1. Volume element $V$ and damage variable **D**.

field [4]. When we need not emphasize the special distribution of the damage fields, they are called *damage variables*. Since these damage variables are macroscopic variables which mechanically describe the state of material deterioration, they are exactly *internal state variables* in the thermodynamic theory of constitutive equations.

According to these assumptions, a rational and systematic approach to the problems of distributed cavities may be established in the framework of continuum mechanics by:

(1) describing the mechanical effects of the distributed cavities in terms of appropriate mechanical variables, i.e. damage variables;

(2) establishing equations which govern the evolution of these damage variables (evolution equations);

(3) formulating equations which describe the mechanical behaviour of damaged materials (constitutive equations);

(4) solving boundary value problems by means of these equations.

Figure 2 shows a comparison between the conventional approach to damage analysis and life prediction, and that of CDM. In the conventional method, the damage state is estimated on the basis of the results of a stress analysis at each stage combined with certain criteria for damage, i.e. it is impossible to incorporate the effect of material damage on the current distribution of stress and strain, or the effect of the current state of damage on the subsequent development of material damage. The CDM approach, on the other hand, can not only pursue the damage process accurately, but also reasonably describe the effect of material damage on the various mechanical properties of a material as well as on the current states of stress and strain.

In the procedure for CDM, furthermore, by introducing a proper criterion of local fracture expressed in terms of a damage variable (together with stress and strain, if necessary), we can describe the initiation of macroscopic cracks as a result of cavity coalescence. Thus, CDM combined with the conventional method of inelastic analysis for boundary value problems furnishes a systematic method to analyze the entire process of damage and fracture, ranging from the material damage due to distributed cavities to the initiation and growth of macroscopic cracks, and finally to the global fracture.

## MODELING OF ANISOTROPIC DAMAGE BY MECHANICAL VARIABLES

The first step in damage and fracture analysis based on CDM is to represent the damage state in terms of appropriate damage variables [12].

The damage in engineering materials usually develops as a result of the nucleation, growth and coalescence of various kinds of microvoids and microcracks at distributed locations in the material. The shape, orientation and distribution of these microcavities, as well as their evolution, usually depend on the direction of the applied stress or applied strain. Thus, the material damage is intrinsically anisotropic, and the anisotropy of material damage sometimes plays an essential role in predicting the fracture process of engineering materials.

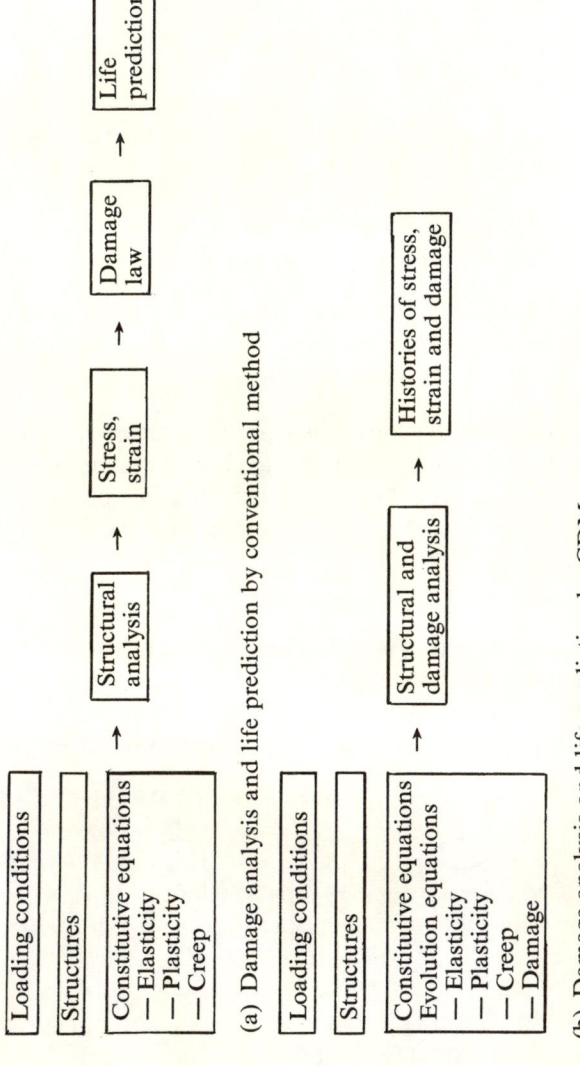

(a) Damage analysis and life prediction by conventional method

(b) Damage analysis and life prediction by CDM

Fig. 2.  Comparison between damage analysis by (a) the conventional method and (b) that of CDM.

Let us consider a bar (Fig. 3) and denote the cross-sectional area of the undamaged initial state and that of the actual damaged state by $A_0$ and $A$, respectively. We assume that the deformation due to the stress $\sigma$ is infinitesimal. (A more elaborate theory based on finite deformation theory was developed recently by Murakami [41].) According to the classical Kachanov–Rabotnov creep damage theory [1, 2, 11], the material damage is identified by the reduction of load-carrying net area due to cavity formation [4], and the damage state is represented by a damage variable $D$; the damage variable $D$ is interpreted to represent the net area reduction and has the values

$$D = 0 \quad \text{(undamaged initial state)}$$
$$D = 1 \quad \text{(final rupture state)} \tag{1}$$

Therefore, the effective net area of the damaged material $A^*$ (Fig. 3(c)) can be represented by

$$A^* = (1 - D)A \tag{2}$$

and the micromechanical state of the current damaged state (Fig. 3(b)) can be assumed to be equivalent to the fictitious undamaged configuration of Fig. 3(c). Then, the net stress $\bar{\sigma}$ of the state (b) magnified by the net area reduction due to damage is equivalent to the stress of state (c):

$$\bar{\sigma} = T/A^* = \sigma/(1 - D) \tag{3}$$

where $T$ is the applied load. Thus, the damage variable $D$ can be defined, once the relationship between the areas $A$ and $A^*$ in these two configurations has been specified.

In order to apply this idea to the three-dimensional state of damage,

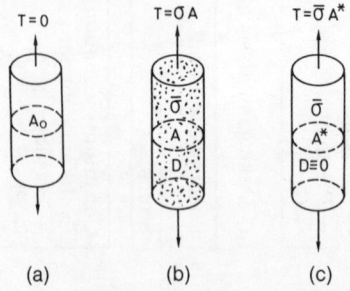

FIG. 3. A bar subject to uniaxial tension (the strain due to the applied stress is assumed to be infinitesimal): (a) initial undamaged state; (b) current damaged state; (c) fictitious undamaged state.

we will first take an area element PQR of an arbitrary orientation in the damaged material as shown in Fig. 4, and call it the *current (actual) damaged configuration* $B_t$. Then, let us represent the line elements PQ, PR and the oriented area of PQR by the vectors dx, dy and $\boldsymbol{v}\,dA$. Because of the three-dimensional distribution of microcavities, the load-carrying net area of PQR will be decreased: we postulate that the net effect is equivalent to a diminished area element P*Q*R* in the fictitious undamaged material, which will be termed the fictitious undamaged configuration $B_f$. The line elements P*Q* and P*R* and the area P*Q*R* will be denoted by the vectors dx*, dy* and $\boldsymbol{v}^*\,dA^*$. Since the net area reduction due to damage occurs not only in the plane PQR but also in planes of other orientations, the directions of the vectors $\boldsymbol{v}\,dA$ and $\boldsymbol{v}^*\,dA^*$ do not always coincide with each other.

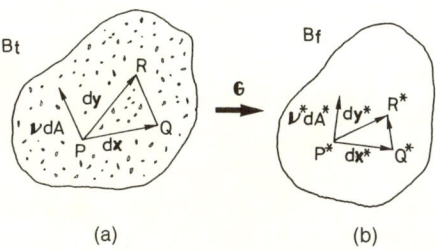

FIG. 4. Definition of the damaged state: (a) current damaged configuration; (b) fictitious undamaged configuration.

The decrease in the net area from PQR to P*Q*R* induced by the cavity distribution can be specified, if we introduce a *fictitious deformation* at the point P from the current configuration $B_t$ of the damaged material to the corresponding fictitious undamaged configuration $B_f$. With the fictitious deformation gradient from $B_t$ to $B_f$ denoted by **G**, the segments dx* and dy* in $B_f$ are given as:

$$dx^* = G\,dx$$

$$dy^* = G\,dy \qquad (4)$$

By the use of Nanson's theorem, the area vector $\boldsymbol{v}^*\,dA^*$ in $B_f$ can be related to the vector $\boldsymbol{v}\,dA$ in $B_t$:

$$\boldsymbol{v}^* dA^* = \tfrac{1}{2} dx^* \otimes dy^* = \tfrac{1}{2}(G\,dx) \otimes (G\,dy) = K(G^{-1})^{T}(\boldsymbol{v}\,dA) \qquad (5)$$

where $K = \det G$ and $(\ )^T$ denotes the transpose of the second-rank tensor.

According to the preceding argument, the damage states can now be specified by the linear transformation $K(\mathbf{G}^{-1})^\mathrm{T}$ of Eqn. (5). Let us represent $K(\mathbf{G}^{-1})^\mathrm{T}$ by an alternative tensor $\mathbf{D}$:

$$K(\mathbf{G}^{-1})^\mathrm{T} = \mathbf{I} - \mathbf{D} \quad \text{or} \quad \mathbf{G} = K[(\mathbf{I} - \mathbf{D})^\mathrm{T}]^{-1} = K[(\mathbf{I} - \mathbf{D})^{-1}]^\mathrm{T} \quad (6)$$

where $\mathbf{I}$ is the unit tensor of rank two and $K = [\det(\mathbf{I} - \mathbf{D})]^{1/2}$. Then, Eqn. (5) has the form

$$\boldsymbol{v}^* \, \mathrm{d}A^* = (\mathbf{I} - \mathbf{D})(\boldsymbol{v} \, \mathrm{d}A) \quad (7)$$

Equation (7) implies that the second-rank tensor $\mathbf{D}$ represents the anisotropic damage states of a material in terms of the net area reduction, and will be called the damage tensor hereafter.

By examining the mechanical properties of the tensor $(\mathbf{I} - \mathbf{D})$ of Eqn. (7), we can assume the symmetry of the tensor $\mathbf{D}$. Then, the damage tensor $\mathbf{D}$ always has three orthogonal principal directions $\mathbf{n}_i$ ($i = 1, 2, 3$) and the corresponding real principal values $D_i$, and can be expressed in the canonical form

$$\mathbf{D} = \sum_{i=1}^{3} D_i \mathbf{n}_i \otimes \mathbf{n}_i \quad (8)$$

In order to examine the property of $\mathbf{D}$, let us take small tetrahedrons OPQR and O*P*Q*R*, which consist of the area elements PQR and P*Q*R* and facets perpendicular to the principal coordinate axes $x_1, x_2$ and $x_3$ as shown in Fig. 5. Substitution of Eqn. (8) into Eqn. (7) furnishes

$$\boldsymbol{v}^* \, \mathrm{d}A^* = \left( \mathbf{I} - \sum_{i=1}^{3} D_i \mathbf{n}_i \otimes \mathbf{n}_i \right)(\boldsymbol{v} \, \mathrm{d}A) = \sum_{i=1}^{3} (1 - D_i) \, \mathbf{n}_i \, \mathrm{d}A_i$$

$$= \mathbf{n}_1 \, \mathrm{d}A_1^* + \mathbf{n}_2 \, \mathrm{d}A_2^* + \mathbf{n}_3 \, \mathrm{d}A_3^* \quad (9a)$$

$$\mathrm{d}A_i^* = (1 - D_i) \, \mathrm{d}A_i \quad \text{(no sum on } i, i = 1, 2, 3) \quad (9b)$$

where $\mathrm{d}A_i = v_i \, \mathrm{d}A$ and $\mathrm{d}A_i^* = v_i^* \, \mathrm{d}A^*$ signify the facet areas of the tetrahedrons in the coordinate planes. According to Eqns. (9), the principal values $D_i$ can be interpreted as the cavity area densities on three principal planes of $\mathbf{D}$ (see Fig. 5).

By using the damage variable $\mathbf{D}$ defined by Eqn. (8) as an internal state variable, the constitutive and evolution equations of the damaged material can be developed by the ordinary theory of inelasticity.

However, the concept of effective stress is often employed in CDM

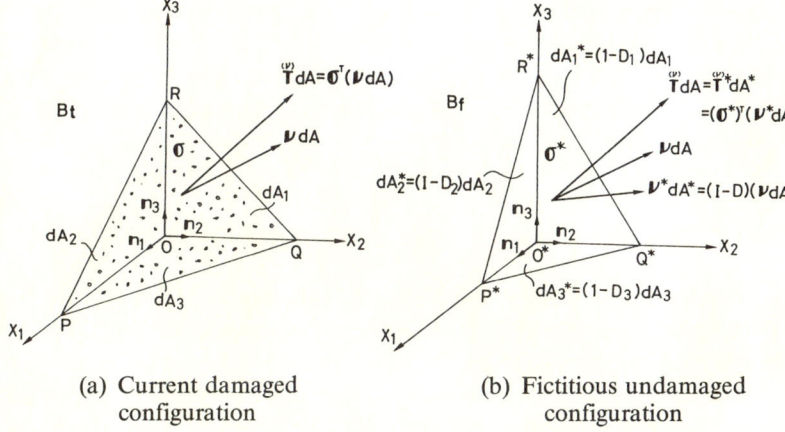

(a) Current damaged configuration

(b) Fictitious undamaged configuration

FIG. 5. Tetrahedrons in the current and the fictitious configuration.

[1–10]. According to the notion of the effective stress expressed in Eqn. (3), the effective stress can be defined as the magnified stress $\sigma^*$ which is induced in the fictitious configuration $B_f$ when the diminished load-carrying area $\nu^*\,dA^*$ is subjected to the same force vector $\mathbf{T}^{(\nu)}\,dA$ as that of the current configuration $B_t$ (Fig. 5). Thus, this stress corresponds exactly to the first Piola–Kirchhoff stress tensor $s$ with respect to the reference configuration $B_f$:

$$\sigma^* = s = K^{-1}G\sigma \qquad (10)$$

Substitution of Eqn. (6) into this equation gives

$$\sigma^* = (K^{-1})K[(I - D)^{-1}]^T\sigma = (I - D)^{-1}\sigma \qquad (11)$$

The effective stress tensor $\sigma^*$ of Eqn. (11) is non-symmetric. Since it is difficult to formulate the evolution equation and the constitutive equation of the damaged material by using a non-symmetric tensor, we need proper symmetrization of this effective stress. One possible procedure for this purpose is to take the symmetric part of the Cartesian decomposition of $\sigma^*$ of Eqn. (11):

$$S = \tfrac{1}{2}[\sigma^* + (\sigma^*)^T] = \tfrac{1}{2}[(I - D)^{-1}\sigma + \sigma(I - D)^{-1}] \qquad (12)$$

The effective stress discussed so far is mainly concerned with that for damage growth. Since the deformation depends not only on the effective area reduction in each plane but also on the three-dimensional cavity distribution, the effective stress for deformation should be defined separately by using the damage tensor **D**. A detailed discussion on this will be found in Ref. 13.

The results of Eqns. (8), (9) and (12) have also been obtained by Murakami and Ohno [14]. The preceding argument, however, gives another and more consistent interpretation of Eqns. (8) and (12).

## CREEP DAMAGE ANALYSIS OF THIN-WALLED TUBES

### Constitutive and Evolution Equations of Creep Damage

As the first example of the application of the damage variable **D** and the effective stress **S** discussed in the preceding section, let us now analyze the process of creep damage and rupture in thin-walled tubes under non-proportional loadings [15]. The damage anisotropy has significant effects on the damage and fracture process, especially in the case where the directions of the principal stress rotate with respect to the material.

We first assume that the creep damage state in polycrystalline metals can be characterized by the three-dimensional cavity area density and can be represented by the second-rank symmetric damage tensor **D** in the form of Eqn. (8). Then, because of the net area reduction due to such a cavity distribution, the effects of the Cauchy stress $\sigma$ are magnified to the effective stress tensor **S** of Eqn. (12).

The rate of damage growth depends mainly on the current states of material damage and stress. Thus, the evolution equation of the damage variable can be specified as follows:

$$\dot{\mathbf{D}} = \mathbf{H}(\mathbf{S}, \mathbf{D}, \kappa) \tag{13}$$

where $\kappa$ is a parameter representing the microstructural changes of the matrix.

Several approaches to materialize the damage equation (13) have been proposed; one is to introduce the damage surfaces in the stress space or in the strain space [16–18], while the other is to establish a dissipation potential on the basis of the thermodynamic theory of inelasticity [19–22]. The tensor function theory [23–26] of non-linear algebra can also be applied to derive the general expressions of Eqn. (13).

However, as an alternative method, we will specify the evolution equation of damage by taking account of the micromechanisms for damage growth and of their mechanical effects. According to the metallurgical observations so far reported, creep cavities develop mainly on grain boundaries perpendicular to the applied stress [27, 28]. These results enable us to represent the increase of the cavity area fraction due to creep damage by a combination of the net area reduction on planes perpendicular to the maximum principal stress direction $v^{(1)}$ and the isotropic area reduction. Thus, in view of the interpretation of the damage tensor $\mathbf{D}$ discussed in the preceding section, the damage rate $\dot{\mathbf{D}}$ can be specified as follows [14]:

$$\dot{\mathbf{D}} = \bar{H}(\mathbf{S}, \mathbf{D}, \kappa)[(1 - \eta)\mathbf{I} + \eta v^{(1)} \otimes v^{(1)}] \qquad (14)$$

where $\bar{H}$ stands for the stress-state dependence of the damage growth rate, and $\eta$ and $\mathbf{I}$ are a material constant and the unit tensor, respectively. By examining the available results of rupture tests, Hayhurst and Leckie [29, 30] proposed an isochronous creep rupture locus in the stress space.

$$\chi(\boldsymbol{\sigma}) = \xi\sigma^{(1)} + \zeta\sigma_{EQ} + \tfrac{1}{3}(1 - \xi - \zeta)\,\mathrm{tr}\,\boldsymbol{\sigma} = 0 \qquad (15)$$

where $\sigma^{(1)}$ and $\sigma_{EQ}$ are the maximum principal stress and the equivalent stress, and $\xi$ and $\zeta$ are material constants. If we assume that the rate of damage growth is characterized by the stress magnitude $\chi(\mathbf{S})$ similar to Eqn. (15), we have an explicit form of Eqn. (14) as follows:

$$\dot{\mathbf{D}} = B[\xi S^{(1)} + \zeta S_{EQ} + \tfrac{1}{3}(1 - \xi - \zeta)\,\mathrm{tr}\,\mathbf{S}]^k[(1 - \eta)\mathbf{I} + \eta v^{(1)} \otimes v^{(1)}] \quad (16)$$

where $S^{(1)}$ is the maximum principal value of $\mathbf{S}$, $S_{EQ} = (\tfrac{3}{2}\,\mathrm{tr}\,\mathbf{S}_D^2)^{1/2}$, $\mathbf{S}_D = \mathbf{S} - \tfrac{1}{3}(\mathrm{tr}\,\mathbf{S})\,\mathbf{I}$, and $B$ and $k$ are material constants.

If we assume that the creep rate $\dot{\boldsymbol{\varepsilon}}^c$ of the damaged material depends on the stress $\boldsymbol{\sigma}$ only through $\mathbf{S}$, the constitutive equation of creep may have the form

$$\dot{\boldsymbol{\varepsilon}}^c = \mathbf{G}(\mathbf{S}, \mathbf{D}, \kappa) \qquad (17)$$

The explicit forms of this equation can be established in several ways e.g. by postulating creep potentials, or by applying the representation theorem for an isotropic tensor function [23–26].

In the case of undamaged materials, a creep analysis of the structural components is often performed by either of the following equations for uniaxial stress:

$$\varepsilon^c = A_1[1 - \exp(-at)]\sigma^{n_1} + A_2\sigma^{n_2}t \qquad \text{(McVetty)} \qquad (18a)$$

$$\varepsilon^c = A_3\sigma^{n_3}t^m \qquad\qquad\qquad \text{(Bailey–Norton)} \qquad (18b)$$

where $\varepsilon^c$ and $t$ stand for creep strain and time, and $a, n_1, n_2, n_3, m, A_1, A_2$ and $A_3$ are material constants.

By postulating a creep flow rule of the von Mises type together with the strain hardening hypothesis, and by using the effective stress of Eqn. (12), Eqns. (18a) and (18b) can be generalized to the creep constitutive equations of damaged materials:

$$\dot{\varepsilon}^c = \tfrac{3}{2}[aA_1(\sigma_{EQ})^{n_1-1}\exp(-a\bar{t})\sigma_D + A_2(S_{EQ})^{n_2-1}\mathbf{S}_D] \tag{19a}$$

$$\varepsilon^c_{EQ}(t) = A_1[\sigma_{EQ}(t)]^{n_1}[1 - \exp(-a\bar{t})] + A_2[\sigma_{EQ}(t)]^{n_2}\bar{t} \tag{19b}$$

$$\varepsilon^c_{EQ}(t) = \int_0^t [\tfrac{2}{3}\operatorname{tr}(\dot{\varepsilon}^c)^2]^{1/2}at \tag{19c}$$

$$S_{EQ} = (\tfrac{3}{2}\operatorname{tr}\mathbf{S}_D^2)^{1/2} \tag{19d}$$

$$\dot{\varepsilon}^c = \tfrac{3}{2}mA_3^{1/m}(\varepsilon_{EQ})^{(m-1)/m}(S_{EQ})^{(n_3-m)/m}\mathbf{S}_D \tag{20}$$

where $\bar{t}$ in Eqn. (19a) is a fictitious time which should be eliminated from Eqn. (19b).

## Creep Damage Analysis of Thin-walled Tubes Under Non-proportional Loading

Let us first discuss a simple problem of creep damage analysis, which was previously analyzed by Kachanov [31] using his vector damage theory. A thin-walled tube is subjected to a constant tensile stress $\sigma$ for the interval $0 < t < t^*$, and subsequently by a constant torsional stress $\tau$ for $t > t^*$. The deformation is assumed to be small. An orthogonal coordinate system $Ox_1x_2$ is taken in the circumferential and the axial direction of the tube.

Let us consider the particular case of $\xi = \eta = 1$ and $\zeta = 0$ in Eqn. (16), i.e.

$$\dot{\mathbf{D}} = B[S^{(1)}]^k(\mathbf{v}^{(1)} \otimes \mathbf{v}^{(1)}) \tag{21}$$

which enables us to obtain an analytical solution for the present problem [15]. Since the tube is subjected to a constant uniaxial tension for $t < t^*$, Eqn. (21) can be analytically integrated to give the damage variable $D^*_{22}$ in the axial direction at $t = t^*$ as follows:

$$D^*_{22} = 1 - [1 - (k+1)B\sigma^k t^*]^{1/(k+1)} \tag{22}$$

We suppose that the rupture takes place under the constant torsional stress $\tau$ for $t > t^*$. Then, the rupture time $t_R$ and the angle $\phi_R$ between the normal to the rupture surface and the $x_1$ axis are obtained as [15]

$$t_R = t^* + [B(k+1)\tau^k]^{-1}[(1-D^*_{22})/(1-\tfrac{1}{2}D^*_{22})]^{k+1} \tag{23}$$

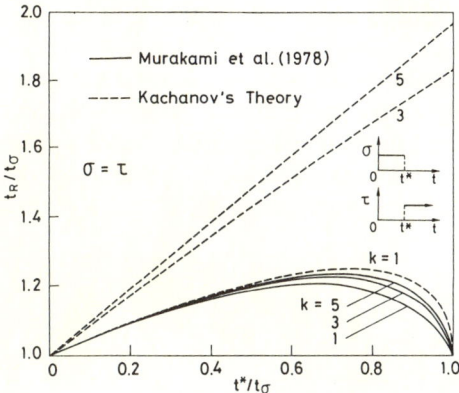

FIG. 6. Creep rupture time for thin-walled tubes under tension followed by torsion.

$$\phi_R = \frac{\pi}{4} + \tfrac{1}{2}\arctan\left[D_{22}^*(2 - D_{22}^*)/\{2(1 - D_{22}^*)\}\right] \qquad (24)$$

Figure 6 shows the relationship between the creep rupture time $t_R$ and the time of stress change $t^*$ in the case of $\sigma = \tau$. In the figure, $t_\sigma$ denotes the rupture time for a uniaxial tension [2, 11, 30]:

$$t_\sigma = [B(k + 1)\sigma^k]^{-1} \qquad (25)$$

The solid lines represent the result of Eqn. (23), while the dashed lines are Kachanov's result [31] calculated from the relationship

$$\dot{\Psi}_v = -B(\sigma_v/\Psi_v)^k \qquad (26)$$

where $\Psi = 1 - D$ and the subscript $v$ denotes a specific plane with unit normal vector $v$.

It can be seen from the figure that the damage process of torsion due to Kachanov's theory for larger values of $k$ proceeds almost independently of the damage accumulated in the preceding process of tension. This is obviously attributable to the simplified assumption in Eqn. (26) that the growth rate of damage on the plane $v$ does not depend on the damage state of other planes.

In order to elucidate the effects of anisotropic creep damage in more detail, Murakami and Ohno [14] further calculated the creep rupture time for thin-walled tubes subjected to combined constant tension and reversed or cyclic torsion (Fig. 7). The material was tough pitch copper at 250 °C, and Eqn. (21) with $B = 1.40 \times 10^{-13}$ MPa$^{-5.6}$ h$^{-1}$ and $k = 5.6$

was employed to describe the evolution of creep damage in this material.

Since this problem permits us no analytical solutions, Eqn. (21) was numerically integrated according to the specified histories of the loading. The curves A and B in Fig. 7 exhibit the rupture time $t_R$ corresponding to the stress histories represented in the figure. The symbol $t_\sigma$ is the rupture time under a constant uniaxial tension of $\sigma = 46\cdot8$ MPa.

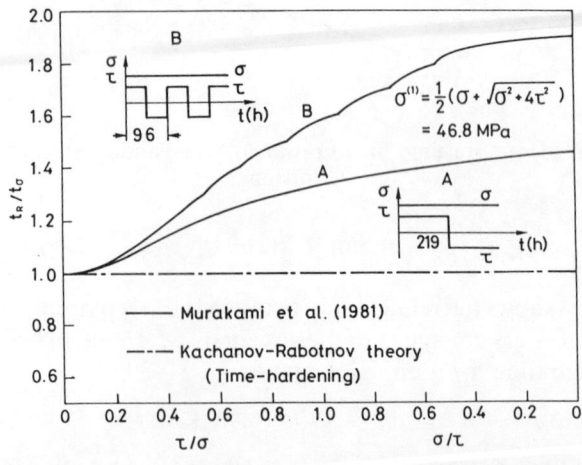

FIG. 7. Effect of material damage on the rupture time of thin-walled tubes under non-proportional loading (curve A, constant tension and reversed torsion; curve B, constant tension and cyclic torsion).

The rupture times for the curves A and B increase with $\tau/\sigma$. This is due to the fact that, when the damage growth is directional, a change in the principal stress direction induces a change in the plane of significant accumulated damage, and thus creep cavities develop mainly on two different planes. This trend has been confirmed by the micrographical observation of Trampczynski et al. [32, 33] for the creep damage of copper under non-proportional loading. On the other hand, the isotropic damage theory represented by a chain line in the figure cannot describe these features; it predicts a constant rupture time $t_R/t_\sigma = 1$ independently of the stress ratio $\tau/\sigma$. It should be noted that an analysis without regard to the damage anisotropy gives dimensionless rupture times from $1/1\cdot5$ to $1/2$ shorter than predictions from the anisotropic theory.

Verification of the anisotropic creep damage theories and their predictions necessitates systematically organized experiments. Murakami et al. [34] performed such tests for creep damage on tough pitch copper tubes at 250 °C under stress histories analogous to those of Fig. 7, and discussed the validity of this analysis. They also refined the analysis by incorporating the effect of creep rates on the damage growth as well as the phenomenon of cross-hardening in creep.

## LOCAL APPROACH TO CREEP CRACK GROWTH UNDER BIAXIAL NON-PROPORTIONAL LOADINGS

CDM describes the local process for material damage. Thus, by introducing an appropriate criterion for local fracture expressed in terms of the damage variable, CDM can simulate the nucleation and growth of discrete cracks, i.e. CDM provides a systematic approach to analyze not only the damage process prior to the initiation of macroscopic cracks, but also the process for the initiation and extension of cracks which have been formed by the coalescence of microscopic cavities.

We will now show the use of CDM for analyzing creep crack growth under complex non-proportional loadings, and discuss the effects of damage anisotropy on crack growth patterns and crack growth rates.

### Finite Element Analysis and Method of Calculation

Equations (16) and (19) were applied to the finite element analysis of creep crack growth in a cracked plate under biaxial loading [12, 42]. The plate was assumed to be in a state of plane stress, and to exhibit elastic–creep deformation with creep incompressibility. By means of the initial strain methods, an analysis was performed on one-quarter of the plate, which was divided into triangular simplex elements as shown in Fig. 8. The numbers of elements and nodes were 396 and 224, respectively. The results of Fig. 8 were also calculated for finer meshes, but the numerical results, especially those for the crack growth pattern, were not much influenced by the division.

The crack growth was simulated by the method employed by Hayhurst et al. [35]: the element was assumed to rupture when the maximum principal value $D_1$ of the damage tensor $\mathbf{D}$ in the element attained $D_1 = 0.99$. Then the stiffness of this element was reduced to zero.

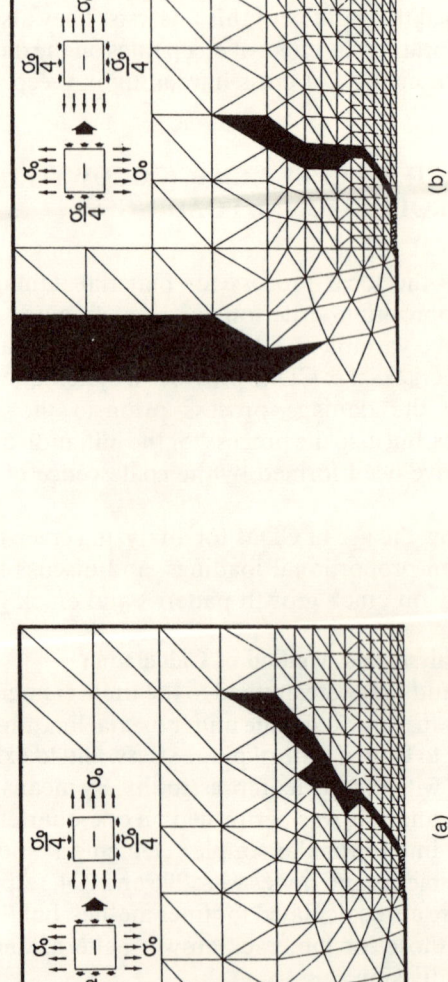

FIG. 8. Creep crack bifurcation in a cracked plate under biaxial non-proportional loading (copper at 250°C; $\sigma_0 = 27.58$ MPa; $a/c = 10$, $b/c = 30$, $2c =$ thickness of plate); (a) isotropic damage ($\eta = 0$; $t_f = 786$ h); (b) anisotropic damage ($\eta = 1.0$; $t_f = 1505$ h).

## Material Constants

The material constants of Eqns. (16) and (19) can be identified by performing a series of constant uniaxial and combined stress tests for copper at 250 °C [36]:

$$B = 4.46 \times 10^{-13} \text{ MPa}^{-k} \text{ h}^{-1},$$

$$k = 5.55, \quad \eta = \begin{cases} 0 & \text{(isotropic)} \\ 1.0 & \text{(anisotropic, plane cracks)} \end{cases}$$

$$\xi = 1.0, \quad \zeta = 0$$

(27a)

$$A_1 = 2.40 \times 10^{-7} \text{ MPa}^{-n_1}, \quad n_1 = 2.60$$

$$A_2 = 3.00 \times 10^{-16} \text{ MPa}^{-n_2} \text{ h}^{-1}, \quad n_2 = 7.10$$

$$a = 0.05 \text{ h}^{-1}$$

(27b)

Young's modulus $E$ and Poisson's ratio $v$ were taken to be equal to those given in Ref. 36; $E = 66\,240$ MPa and $v = 0.3$.

## Results of Analysis and Discussion

Figure 8 shows the results from a finite element analysis of a cracked copper plate at 250 °C under the biaxial non-proportional stress represented in the figure. In this figure, the first load was maintained until the eight elements in front of the initial crack were ruptured, and then the load was changed to the second value. In either of Figs. 8(a) and 8(b), the incipient rupture at the first element in front of the crack tip occurred at $t = 24$ h, whereas the time for the eighth element to rupture (i.e. the time of the stress change) was 95 h; for proportional loading, the isotropic and anisotropic damage theories gave almost identical results.

After the stress change, the crack of Fig. 8(a) (isotropic, $\eta = 0$) changed its direction and then developed in that direction until the final rupture at $t_f = 786$ h. In Fig. 8(b) (anisotropic, $\eta = 1$), on the other hand, the crack first propagated to the right and upwards for a while, and then changed to a direction parallel to the $y$ axis. The final fracture occurred at $t_f = 1505$ h by the coalescence of this crack and the newly developed crack on the $y$ axis. Namely, the anisotropic damage theory predicted creep rupture times about twice as long as those from the isotropic theory. Thus, the anisotropy of creep damage has a significant effect not only on the crack propagation patterns, but also on the final rupture times.

## EFFECTS OF PRIOR PLASTIC DAMAGE ON THE SUBSEQUENT PROCESS OF CREEP RUPTURE

The continuum damage theory developed so far is also applicable to other types of damage, including creep–fatigue damage, fatigue damage and elastic–plastic damage.

As already mentioned, material damage in general has a significant influence on various mechanical and physical properties of the material. According to the observations on some nickel-base alloys [37], for example, distributed macroscopic cavities brought about by plastic deformation at room temperature largely cause the subsequent creep strength to deteriorate at elevated temperatures, and their effects have marked anisotropy. Let us show that CDM can be applied successfully also to this problem [38].

The plastic damage of polycrystalline metals in the temperature range of significant slip or grain boundary sliding usually develops as a result of cavity formation on the grain boundary [33]. As we have already discussed, the creep damage is also governed by the development of grain boundary cavities. Thus, if we postulate that the damage state of plastic damage can be characterized by the second-rank symmetric tensor of Eqn. (8), the damage rate $\overset{o}{\mathbf{D}}$ may be decomposed into the sum of the plastic rate $\overset{o}{\mathbf{D}}{}^{p}$ and the creep damage rate $\overset{o}{\mathbf{D}}{}^{c}$:

$$\overset{o}{\mathbf{D}} = \overset{o}{\mathbf{D}}{}^{p} + \overset{o}{\mathbf{D}}{}^{c} \qquad (28)$$

where ($\overset{o}{\ }$) denotes the Jaumann derivative.

Dyson et al. [37] applied plastic prestrains in tension, compression and torsion to Nimonic 80A specimens at room temperature, annealed them for 2 h at 750°C, and then quantitatively observed the cavities formed in the specimens under a high-voltage electron microscope. The characteristic features of this observation were as follows.

(1) Cavities due to plastic damage were observed mainly on grain boundaries parallel to the direction of maximum principal stress.

(2) The volume density of the grain boundary cavities increased with the equivalent plastic strain.

(3) The magnitude of the cavity density was larger for tensile prestrains than that for torsional prestrains of the same value of equivalent plastic strain.

A simple evolution equation for the plastic damage conforming to these observations is

$$\overset{\circ}{\mathbf{D}}{}^{\mathrm{p}} = C\{(K_2^3 + \lambda K_3^2)/[1 + (\tfrac{4}{27})\lambda]\}^l[(\tfrac{2}{3}) \operatorname{tr}(\mathbf{d}^{\mathrm{p}})^2]^{1/2}[2\mathbf{I} - \mathbf{v}^{\mathrm{p}(1)} \otimes \mathbf{v}^{\mathrm{p}(1)}]$$

$$(29a)$$

$$K_2 = \tfrac{1}{2}\int_0^t \operatorname{tr}[\mathbf{d}^{\mathrm{p}}(\tau)]^2 \, \mathrm{d}\tau$$

$$(29b)$$

$$K_3 = \tfrac{1}{3}\int_0^t \operatorname{tr}[\mathbf{d}^{\mathrm{p}}(\tau)]^3 \, \mathrm{d}\tau$$

$$(29b)$$

where $\mathbf{v}^{\mathrm{p}(1)}$ is the unit vector in the direction of the maximum positive principal value of the plastic deformation rate tensor $\mathbf{d}^{\mathrm{p}}$, and $C, l$ and $\lambda$ are material constants.

As regards the evolution equation for creep damage of Nimonic 80A at elevated temperatures, we assume that Eqn. (16) with $\zeta = 1 - \xi$ and $\eta = 1$ holds:

$$\overset{\circ}{\mathbf{D}}{}^{\mathrm{c}} = B[\xi S^{(1)} + (1 - \xi)S_{\mathrm{EQ}}]^k(\mathbf{v}^{\mathrm{c}(1)} \otimes \mathbf{v}^{\mathrm{c}(1)})$$

$$(30)$$

The creep curves of this material under constant tension and under constant torsion have no apparent primary creep stage, but show a secondary creep stage from the beginning and then lead to a tertiary stage. As regards this tertiary stage, besides the acceleration induced by the cavity formation as a result of creep damage, material softening due to the dislocation networks at particle interfaces has been pointed out [39, 40]. If we assume that the change in such dislocation structures in the tertiary creep can be represented by the scalar internal state variable $R$, the creep constitutive equation of this material can be represented as follows:

$$\mathbf{d}^{\mathrm{c}} = \tfrac{3}{2}A[S_{\mathrm{EQ}}/R]^{n-1}(\mathbf{S}_{\mathrm{D}}/R)$$

$$(31a)$$

$$\dot{R} = P(q - R)[\tfrac{2}{3}\operatorname{tr}(\mathbf{d}^{\mathrm{c}})^2]^{1/2}$$

$$(31b)$$

where $A, n, P$ and $q$ are material constants.

By using the above equations, the effects of prior plastic prestrains at room temperature on the subsequent creep damage process for Nimonic 80A at 750 °C were analyzed in Ref. 38.

Figure 9 shows the torsional creep curves for Nimonic 80A at 750 °C that was subjected to prior plastic torsion. The solid and dotted lines in the figure are the numerical predictions of Eqns. (28)–(31) and the

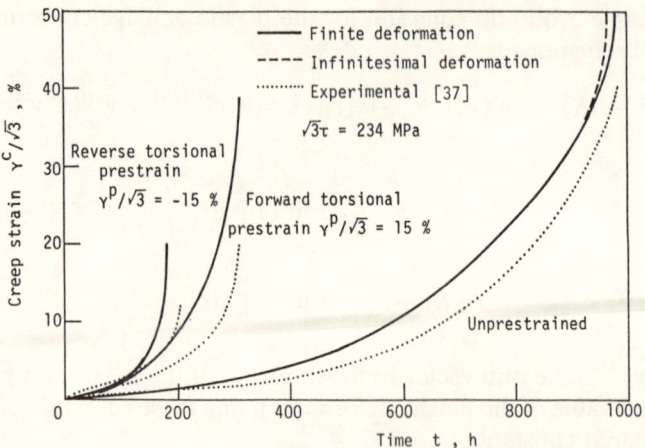

FIG. 9. Torsion creep curves after plastic pretorsion at room temperature (Nimonic 80A; 750°C; $\sqrt{3}\tau = 234$ MPa).

corresponding experimental results by Dyson *et al.* [37], while the dashed line is the prediction of the infinitesimal deformation theory. The creep curve for unprestrained material and the relevant numerical results are also presented in the figure for reference. Similar results were also observed for creep curves under different levels of torsional stress as well as for tensile creep tests after plastic pretension.

One of the significant aspects that can be observed in this figure is the marked dependence of the subsequent creep behavior on the direction of the prior plastic torsion. Such differences in creep rupture times are attributable to the anisotropic features of the plastic and creep damage, and cannot be accounted for by the isotropic damage theories reported hitherto. It can be confirmed from this figure that the continuum damage theory developed in this paper closely describes the experimental results.

## CONCLUDING REMARKS

The process for crack growth to final fracture at high temperatures is usually preceded by the formation of microscopic voids and cavities, and is associated with inelastic and dissipative local deformation. As a systematic approach to these problems that is an alternative to the conventional method for life prediction, the possibility and the applicability of CDM were discussed in some detail. Although the relevant problems in this chapter were mainly creep damage and plastic damage, this approach is also applicable to fatigue and creep–fatigue problems. Since the initiation and growth of fatigue cracks are directional, the present anisotropic damage theory will be effective in these problems, and they will be the objectives of future work by the present authors.

## REFERENCES

[1]   L. M. Kachanov, *Izv. Acad. Nauk SSSR, Otd. Tekh. Nauk,* **8**, 26 (1958) (in Russian).
[2]   L. M. Kachanov, *Introduction to Continuum Damage Mechanics,* Martinus Nijhoff (1986).
[3]   J. Janson and J. Hult, *J. Méc. Appl.*, **1**, 69 (1977).
[4]   J. Hult, in *Mechanisms of Deformation and Fracture*, K. E. Easterling, Ed., 233, Pergamon, Oxford (1979).
[5]   J. Lemaitre and J. L. Chaboche, *J. Méc. Appl.*, **2**, 317 (1978).
[6]   J. Lemaitre, *Trans. 5th Int. Conf. Struct. Mech. Reactor Tech.*, North-Holland, Amsterdam, Paper No. L5/1*b (1979).
[7]   J. L. Chaboche, *Nucl. Eng. Des.*, **64**, 233 (1981).
[8]   J. L. Chaboche, *Nucl. Eng. Des.*, **79**, 309 (1984).
[9]   D. Krajcinovic, *Appl. Mech. Rev.*, **37**, 1 (1984).
[10]  S. Murakami, *J. Eng. Mater. Technol., Trans. ASME,* **105**, 99 (1983).
[11]  Yu. N. Rabotnov, *Creep Problems in Structural Members*, North-Holland, Amsterdam (1969).
[12]  S. Murakami, *Proc. 2nd Int. Conf. on Constitutive Laws Eng. Mater.*, C. S. Desai and E. Krempl, Eds., Elsevier Applied Science, Barking (1987).
[13]  S. Murakami and T. Imaizumi, *J. Méc. Théor. Appl.*, **1**, 743 (1982).
[14]  S. Murakami and N. Ohno, *Creep in Structures*, A. R. S. Ponter and D. R. Hayhurst, Eds., 422, Springer, Berlin (1981).
[15]  S. Murakami and N. Ohno, *Mem. Fac. Eng., Nagoya Univ.*, **36**, 179 (1984).
[16]  J. W. Dougill, *Q. J. Appl. Math.*, **33**, 233 (1975).
[17]  A. Dragon and Z. Mróz, *Int. J. Eng. Sci.*, **17**, 121 (1979).
[18]  D. Krajcinovic and G. U. Fonseka, *J. Appl. Mech., Trans. ASME,* **48**, 809 (1981).
[19]  J. Lemaitre and J. L. Chaboche, *Mécanique Matériaux des Solides*, Dunod, Paris (1985).
[20]  J. P. Cordebois and F. Sidoroff, *J. Méc. Théor. Appl.*, Numéro Spécial, 45 (1982).
[21]  D. Krajcinovic, *J. Appl. Mech., Trans. ASME,* **50**, 355 (1983).
[22]  D. Krajcinovic and S. Selvaraj, *J. Eng. Mater. Tech., Trans. ASME,* **106**, 405 (1984).
[23]  C. Truesdell and W. Noll, *The Non-Linear Field Theories of Mechanics*, Vol. III/3, *Encyclopedia of Physics*, S. Flügge, Ed., Springer, Berlin (1965).

[24] D. C. Leigh, *Nonlinear Continuum Mechanics*, McGraw-Hill, New York (1968).

[25] A. J. M. Spencer, *Theory of invariants*, Vol. I, *Continuum Physics*, A. C. Eringen, Ed., Academic Press, New York (1971).

[26] J. Betten, *J. Méc. Théor. Appl.*, **2**, 13 (1983).

[27] F. Garofalo, *Fundamentals of Creep and Creep-Rupture*, Macmillan, New York (1965).

[28] H. E. Evans, *Mechanisms of Creep Fractures*, Elsevier Applied Science, Barking (1984).

[29] D. R. Hayhurst, *J. Mech. Phys. Solids*, **20**, 381, (1972).

[30] F. A. Leckie and D. R. Hayhurst, *Acta Metall.*, **25**, 1059 (1977).

[31] L. M. Kachanov, *Foundations of Fracture Mechanics*, Nauka, Moscow (1974) (in Russian).

[32] W. A. Trampczynski, D. R. Hayhurst and F. A. Leckie, *J. Mech. Phys. Solids*, **29**, 353 (1981).

[33] S. Murakami, in *Failure Criteria of Structured Media*, J. P. Boehler, Ed., A. A. Balkema, Rotterdam (1986).

[34] S. Murakami, Y. Samomura and K. Saitoh, *J. Eng. Mater. Technol., Trans. ASME*, **108**, 167 (1986).

[35] D. R. Hayhurst, P. R. Brown and C. J. Morrison, *Philos. Trans. R. Soc. London, Ser. A*, **311**, 131 (1984).

[36] D. R. Hayhurst, P. R. Brown and M. W. Chernuka, *J. Mech. Phys. Solids*, **23**, 335 (1975).

[37] B. F. Dyson, M. S. Loveday and M. J. Rodgers, *Proc. R. Soc. London, Ser. A*, **349**, 245 (1976).

[38] S. Murakami and Y. Sanomura, *Eng. Fracture Mech.*, **25**, 693 (1986).

[39] M. F. Ashby and B. F. Dyson, *Proc. 6th Int. Conf. on Fracture, 1984*.

[40] P. J. Henderson and M. McLean, *Acta Metall.*, **31**, 1203 (1983).

[41] S. Murakami, *J. Appl. Mech., Trans. ASME*, **55** (1988) (in press).

[42] S. Murakami, M. Kawai and H. Rong, *Int. J. Mech. Sci.*, **30** (1988) (in press).

# Characterization of High Temperature Strength of Metals Based on the Mechanics of Crack Propagation

RYUICHI OHTANI and TAKAYUKI KITAMURA
*Department of Engineering Science, Faculty of Engineering, Kyoto University, Yoshida, Sakyo-ku, Kyoto 606, Japan*

## ABSTRACT

Metallic materials show different types of strength behavior according to the operating conditions of stress, temperature and time, and certain typical types of behavior are usually distinguished and designated as different phenomena. The crack propagation behavior of such phenomena was investigated on the basis of fracture mechanics. The results are summarized and the characterization of high temperature strength is given in this paper. The crack propagation at high temperatures is clearly divided into two types: time dependent and cycle dependent. The crack propagation rate is correlated well with the creep $J$-integral range for the former and with the fatigue $J$-integral range for the latter. The boundary between the two types is given as a function of the $J$-integral ranges. Moreover, the fatigue life laws of a smooth specimen at high temperatures can be derived from the crack propagation laws, because the small cracks initiate at a very early stage of the life and the failure is subject to their propagation. A similar characterization to that for the crack propagation can be expected for the life laws of a smooth specimen.

## INTRODUCTION

The strength of metallic materials at high temperatures has been phenomenally labeled as monotonic creep, dynamic creep, cyclic creep, tensile strength, low cycle fatigue, high cycle fatigue, thermal fatigue etc., according to the conditions of stress or strain cycle, temperature and time. Not only their characteristics but also the differences between them, however, are still vague. The complexity mainly originates from creep, which has a time effect on the other strengths. The effects of the environment and the microstructural change in materials, which have other time dependences, are not referred to in this chapter. Figure 1 schematically illustrates the effects of stress cycle, cycle frequency, temperature and creep on the strength phenomena in three-dimensional coordinates of mean stress, stress amplitude and temperature. A brief outline of the phenomena is given as follows.

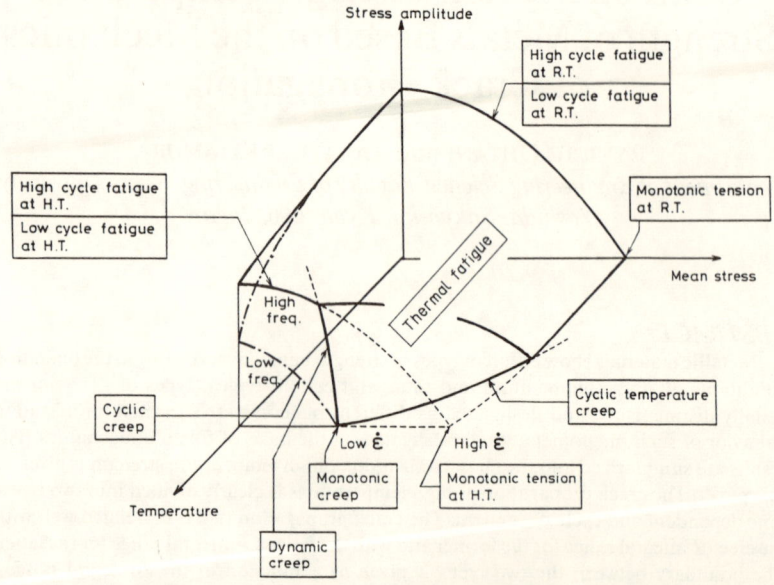

FIG. 1. Schematic diagram explaining the classification of high temperature strength on the basis of temperature, mean stress and stress amplitude.

The monotonic deformation is usually divided into two types, namely 'tensile deformation' and 'creep deformation', and the difference is their time dependence. Let us assume an ideal tensile deformation being purely time independent, which is expected to take place when the strain rate is fast enough to avoid the creep effect even at high temperatures. On the other hand, the creep deformation is defined as purely time dependent under a low strain rate. The usual tensile tests, however, are conducted at a medium strain rate and the test results are midway between ideal tension and creep.

Creep with a stress change is classified into 'dynamic creep' and 'cyclic creep' according to the stress waveform. The former means creep under a high frequency vibration stress of a rather small amplitude superposed on a high mean stress. The latter is creep under a low frequency periodical change of stress. The typical stress waveforms are shown in Fig. 2. In some cases, the stress change drastically accelerates the creep rate and the crack propagation rate, leading to shortening of the creep life.

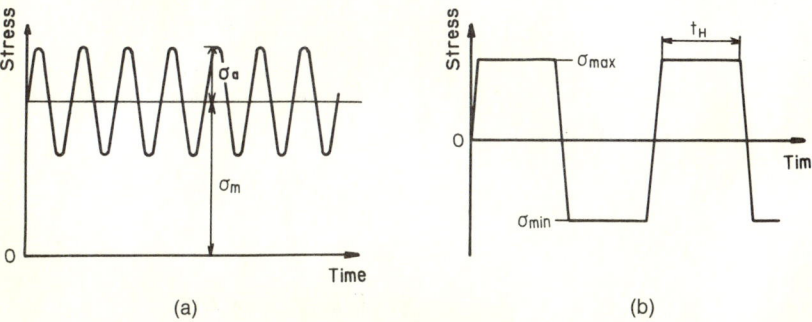

FIG. 2. Stress waveforms for dynamic creep and cyclic creep: (a) dynamic creep; (b) cyclic creep.

'Isothermal fatigue' is often divided into 'high cycle fatigue' and 'low cycle fatigue', for which the difference is originally defined on the basis of the number of cycles to fatigue failure. The definition, however, has no physical meaning. It is rather logical to separate them into elastic fatigue and elastic–plastic fatigue, or small-scale yielding fatigue and large-scale yielding fatigue, because linear fracture mechanics is applicable to the former but not to the latter. For high temperature fatigue, the creep effect is the most important problem, and will be discussed in detail in this chapter.

The temperature effect couples closely with the creep effect as the creep phenomenon is a thermally activated process. The fatigue strength at low temperatures (e.g. at room temperature), therefore, can be expressed by only one curve on the stress amplitude versus mean stress plane in Fig. 1, while that at high temperatures has to be given by many different curves for different stress frequencies and waveforms.

'Thermal fatigue' sometimes means the fatigue under a pure thermal stress cycle, and it is called 'thermal–mechanical fatigue' for fatigue under the combination of thermal stress and mechanical stress cycles. In a wide sense, however, 'thermal fatigue' includes all fatigue

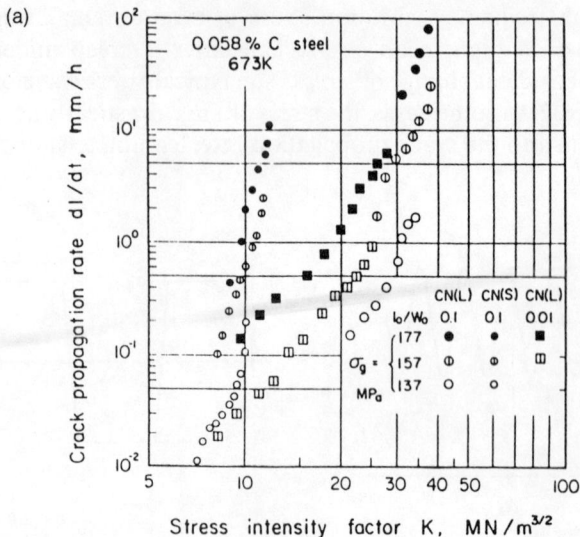

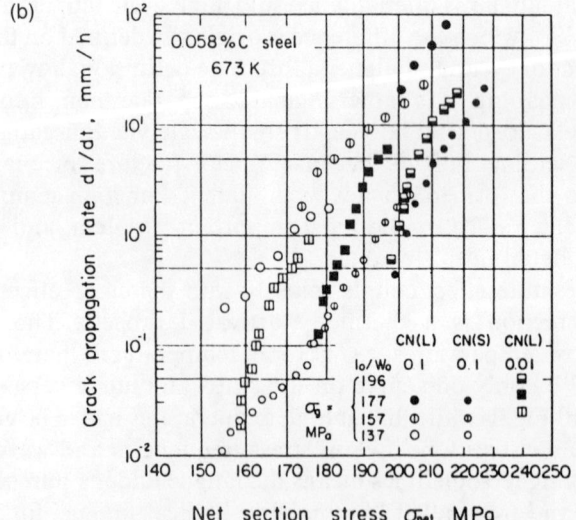

FIG. 3. Relationships between the crack propagation rate $\mathrm{d}l/\mathrm{d}t$ and the fracture mechanics parameters in monotonic creep: (a) $\mathrm{d}l/\mathrm{d}t$ versus $K$ relationship; (b) $\mathrm{d}l/\mathrm{d}t$ versus $\sigma_{\mathrm{net}}$ relationship.

(c)

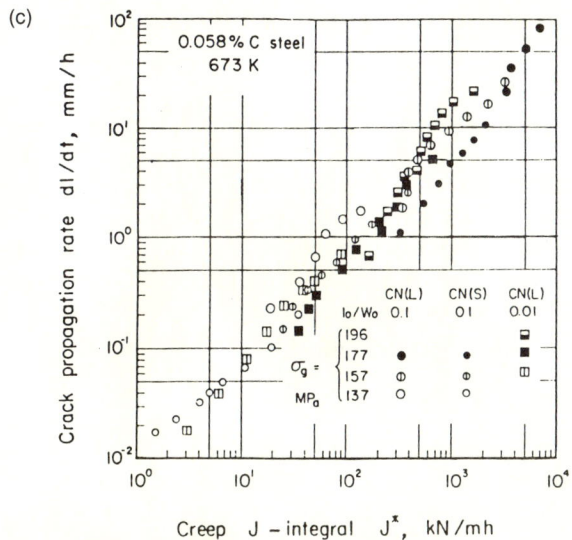

FIG. 3. — *contd.* (c) d*l*/d*t* versus *J**\* relationship.

phenomena under cyclic temperature conditions. Creep or fatigue subjected to both cyclic stress and cyclic temperature, therefore, can be placed in the category of thermal fatigue.

The crack propagation behavior under creep and fatigue conditions has been studied by the authors, and the crack propagation laws for each strength phenomenon have been clarified [1-22]. In this chapter the results are reviewed and summarized in order to characterize the phenomena and to give an overall view of high temperature strength, especially on creep and fatigue and their interaction.

## CRACK PROPAGATION UNDER MONOTONIC CREEP

The fracture mechanics parameters usually used for crack propagation under monotonic creep are the elastic stress intensity factor $K$, the net section stress $\sigma_{net}$ and the creep $J$-integral $J^*$ (modified $J$-integral [23], $C^*$-parameter [24-26]). Figure 3 shows the relationships between the crack propagation d$l$/d$t$ and these parameters in a thin plate specimen of 0·058% carbon steel tested at 673 K (400 °C) [4, 7]. Here, the initial plate widths $2W_0$ of the specimens of type CN(L) and CN(S) are 160 mm and 16 mm, respectively, and $l_0$ is the initial half crack length. $J^*$ is evaluated from the following equation proposed by Ohji *et al.* [27]:

$$J^* = \frac{n-1}{n+1}\sigma_{net}\dot{V}_c \tag{1}$$

where $n$ is the stress exponent of the power law creep and $\dot{V}_c$ is the crack center opening displacement rate. The $dl/dt$–$K$ and the $dl/dt$–$\sigma_{net}$ relationships are strongly dependent on the specimen width, but the $dl/dt$–$J^*$ relationship is independent of it. The results prove the validity of the creep $J$-integral as the controlling parameter of the creep crack propagation rate. The relationship is formulated as

$$dl/dt = C_c J^{*m_c} \tag{2}$$

where $C_c$ and $m_c$ are the material constants.

Figure 4 shows the $dl/dt$–$J^*$ relationship in monotonic creep of metals, obtained up to now by the group of the authors [1, 14]. It reveals that the difference in the $dl/dt$–$J^*$ relationship between the materials is small and that $m_c$ is nearly equal to unity for all materials. The authors have already studied and reported elsewhere the effect of various factors on the creep crack propagation such as the environment [1], temperature [1], biaxial stress [6], fracture mode [1], pre-creep strain [1] and others [10].

## CRACK PROPAGATION DURING STRESS RELAXATION

The creep $J$-integral $J^*$ in a creeping body is analogous to the $J$-integral $J$ in a linear or nonlinear elastic body. The creep $J$-integral is defined as the path-independent line integral converting the strain and the displacement in the $J$-integral equation into the strain rate and the displacement rate, respectively [23, 24]. Because the $J$-integral $J$ can be defined even in an unloading condition for an elastic body (although it cannot for a plastic body), the creep $J$-integral $J^*$ can be expected to be valid in a decreasing load condition, so long as the constitutive relationship of stress and strain rate can be expressed by the following type of equation at any moment:

$$\dot{\varepsilon} = f(t)g(\sigma) \tag{3}$$

where $f(t)$ and $g(\sigma)$ are functions of time and stress.

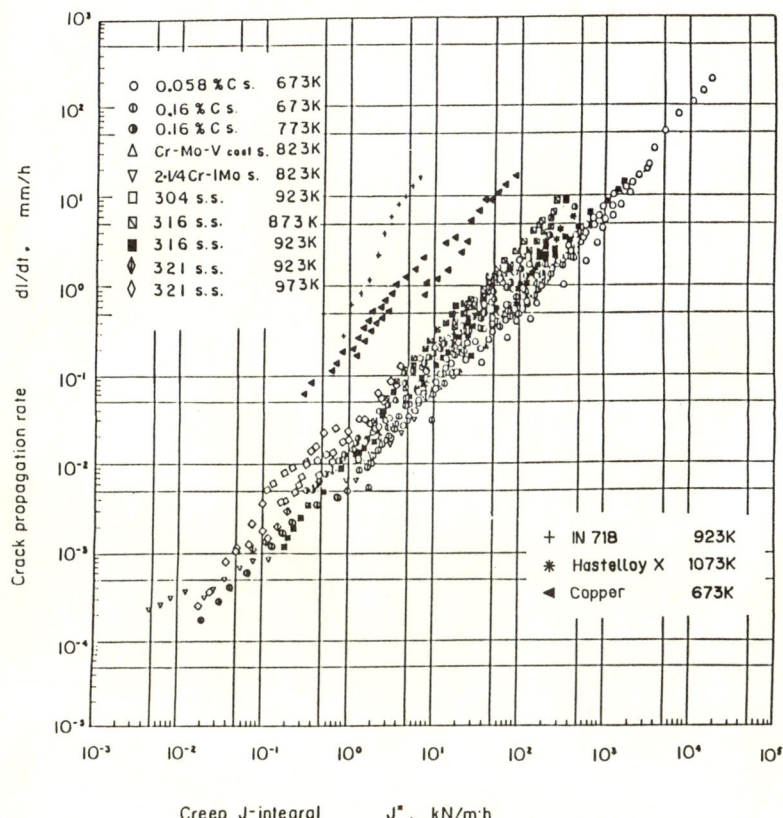

FIG. 4. Relationship between $dl/dt$ and $J^*$ in monotonic creep for several kinds of metallic materials.

The authors conducted creep crack propagation tests under stress relaxation conditions to verify the $J^*$-integral approach in the strain hold period of low cycle fatigue [21]. Figure 5 shows the typical results, indicating that the same straight-line relationship can be obtained even when the crack propagation rate decreases with the decrease in magnitude of $J^*$ by the stress relaxation.

## CRACK PROPAGATION UNDER DYNAMIC CREEP

Figure 6 shows the dependence of the crack propagation rate $dl/dt$ on the stress frequency $v$ and the stress ratio $R$ in the dynamic creep of a

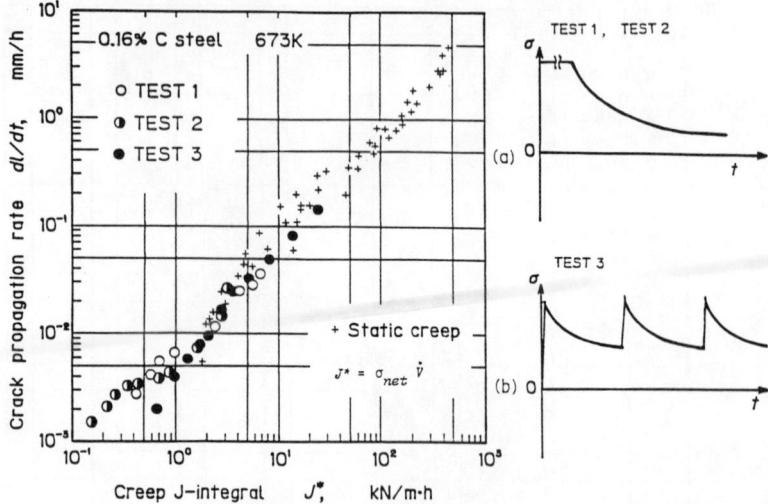

FIG. 5. Relationship between the crack propagation rate $dl/dt$ and creep $J$-integral $J^*$ in stress relaxation.

type 304 stainless steel tested at 923 K (650 °C) [9]. It becomes clear from the figure that $dl/dt$ depends little on the frequency and increases as the equivalent stress $\sigma_e$ increases. Here, $\sigma_e$ is defined as the time-averaged stress in a cycle and is calculated by [28]

$$\sigma_e = \sigma_m \left[ \int_0^1 \left\{ 1 + \frac{\sigma_a}{\sigma_m} \sin(2\pi\nu t) \right\}^n d(\nu t) \right]^{1/n} \qquad (4)$$

where $\nu$ is the frequency, $\sigma_a$ is the stress amplitude and $\sigma_m$ is the mean stress.

Figure 7 shows the relationship between $dl/dt$ and $J^*$ of the above tests except for $R = -0.50$. $dl/dt$ has a fairly good correlation with $J^*$ and the relation corresponds to that of monotonic creep. Moreover, intergranular facets were observed on the fracture surface except for $R = -0.50$. This result also indicates that the crack propagated by creep.

Striations, which suggest fatigue crack propagation, were observed on the fracture surface for $R = -0.50$, so that the mutual relationship between dynamic creep and fatigue comes into this problem. This will be discussed later in detail.

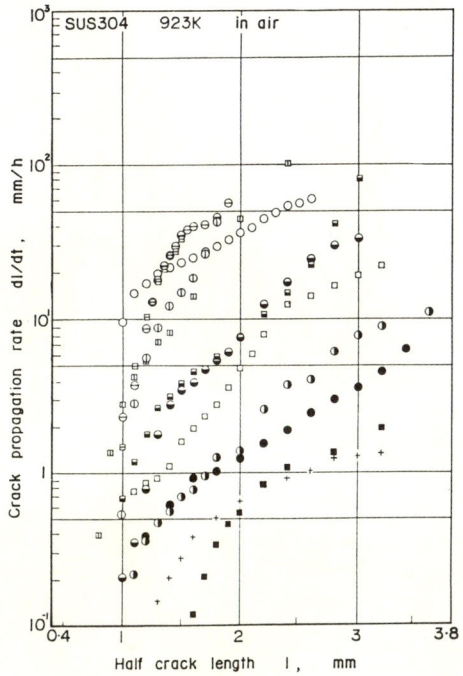

| | Stress ratio $R = \sigma_{min}/\sigma_{max}$ | Mean stress $\sigma_m$ (MPa) | Stress amplitude $a_a$ (MPa) | Frequency $\nu$ (Hz) | Symbol in figures |
|---|---|---|---|---|---|
| Static creep | 1 | 137·2 | 0 | | + |
| Dynamic creep | 0·87 | 137·2 | 9·8 | 5 0·5 | ● ■ |
| | 0·75 | 137·2 | 19·6 | 5 | ◐ |
| | 0·50 | 137·2 | 45·7 | 5 0·5 | ◓ ▤ |
| | 0·33 | 137·2 | 68·6 | 5 0·5 | ◫ ⊞ |
| | 0 | 102·9 | 120·9 | 5 0·5 | ◒ ⊟ |
| | −0·50 | 44·1 | 132·3 | 3 0·3 | ○ □ |

FIG. 6. Dependence of the crack propagation rate on the frequency and stress ratio in dynamic creep.

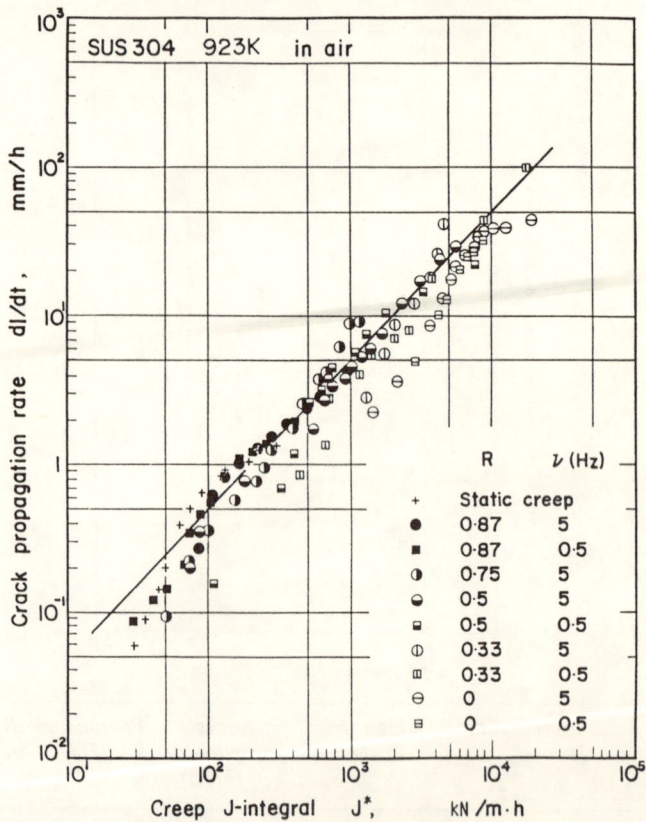

FIG. 7. Relationship between the crack propagation rate $dl/dt$ and the creep $J$-integral $J^*$.

## CRACK PROPAGATION UNDER CYCLIC CREEP

Figure 8 shows the crack propagation rate in cyclic creep of 0·16% carbon steel tested at 673 K (400 °C) using a trapezoidal stress waveform (Fig. 2(b)) with a 1 min stress hold time [5, 7]. As the stress ratio $R$ decreases, the crack propagation rate increases markedly. The creep rate near the crack tip $\dot{\varepsilon}_{ct}$ during the tension period was measured simultaneously by the elongation of a nickel grid, whose pitch was 50 $\mu$m and was secured on the specimen surface by baking before the tests. The crack tip creep rate $\dot{\varepsilon}_{ct}$ was also accelerated in the lower stress ratio tests as shown in Fig. 9, this acceleration of creep rate being caused by dynamic recovery during the compression period[†]. $J^*$ is the

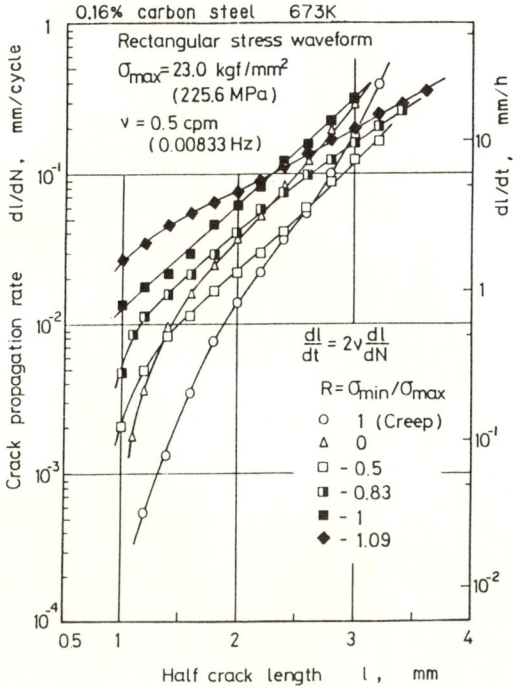

FIG. 8. Dependence of the crack propagation rate on the stress ratio in cyclic creep.

parameter which represents the intensity of creep rate in the vicinity of the crack tip and can describe the acceleration of $\dot{\varepsilon}_{ct}$ caused by the dynamic recovery and/or the small-scale creep. Therefore, $dl/dt$ has a good correlation with $J^*$ for cyclic creep as shown in Fig. 10.

Carrying through the time integral of Eqn. (2) during a cycle [3],

$$dl/dN = C_c \, \Delta J_c \tag{5}$$

$$\Delta J_c = \int J^* \, dt \tag{6}$$

are obtained for $m_c = 1$ in Eqn. (2). Here, $dl/dN$ is the crack propagation

†Another acceleration mechanism in cyclic creep is small-scale creep [29, 30], which is the phenomenon whereby the creep strain grows preferentially near the crack tip due to the constraint of the surrounding elastic strain field. This becomes a serious problem with high strength materials and under low stress conditions [12, 13]. The effect, however, is small enough in this case [17].

# R. OHTANI AND T. KITAMURA

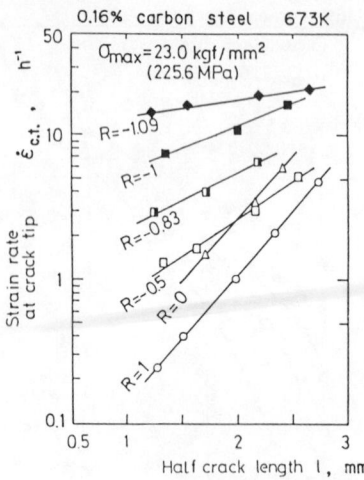

FIG. 9. Strain rate near the crack tip in cyclic creep.

rate (the crack propagation per unit cycle) and $\Delta J_c$ is the time integral of $J^*$ (named the creep $J$-integral range). The expression is sometimes more convenient for creep crack propagation under cyclic stress conditions. The value of $\Delta J_c$ was simply evaluated by a load versus crack center opening displacement ($P$-$V$) loop as shown in Fig. 11(a), which was derived from the integral of Eqn. (1) [3, 8]. Supposing that the inelastic strain is composed of the creep (time-dependent) strain and the plastic (time-independent) strain, the method can be extended for the general stress (or strain) waveforms as illustrated in Fig. 11(b) [3, 8]. In this method, the inelastic strain can be partitioned in a similar way to the strain partitioning method (rapid straining method) [31, 32].

The authors conducted strain-controlled tests using a triangular strain waveform with low frequency and verified the validity of $\Delta J_c$. As a result, $\Delta J_c$ was a reasonable parameter for $dl/dN$ independently of the stress (or strain) waveform when the crack propagated by creep. Figure 12 shows the $dl/dN$–$\Delta J_c$ relationships of several kinds of steels and alloys under various stress and temperature conditions [14].

## CRACK PROPAGATION UNDER HIGH CYCLE FATIGUE

Figure 13 shows the relationship between $dl/dN$ and the stress intensity factor range $\Delta K$ for 0·16% carbon steel tested at 673 K (400 °C) in high frequency (1–3 Hz) fatigue with a stress ratio of −1. The dotted

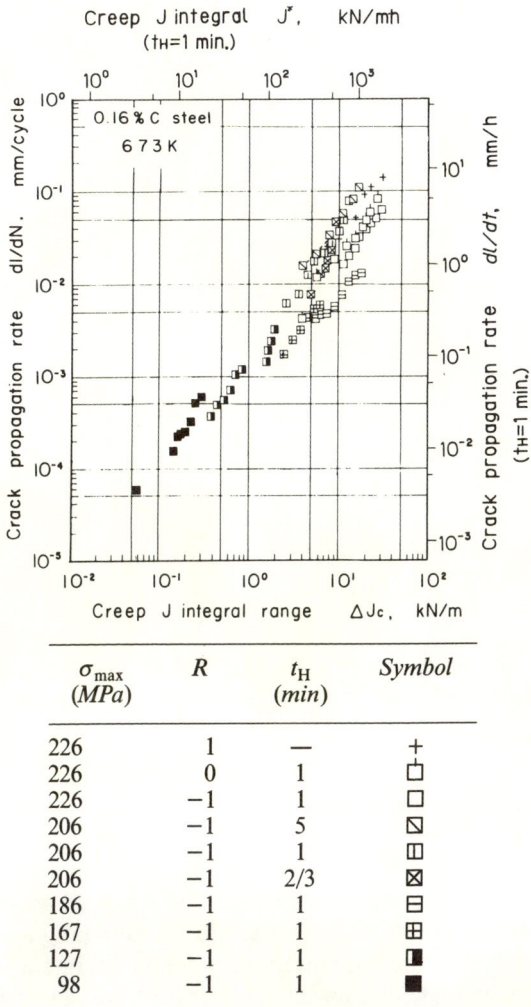

FIG. 10. Relationship between d$l$/d$t$ and $J^*$ in cyclic creep.

line shows the relationship between d$l$/d$N$ and the effective stress
intensity factor range $\Delta K_{\text{eff}}$ which is defined as the difference between
the maximum and the crack tip opening stress intensity factors. The
figure reveals the existence of the crack propagation threshold in fatigue
and good correlations under the lower stress conditions. The propagation
curves have a strong resemblance to those for room temperature fatigue.
Moreover, striations, which suggest that the crack propagated owing to

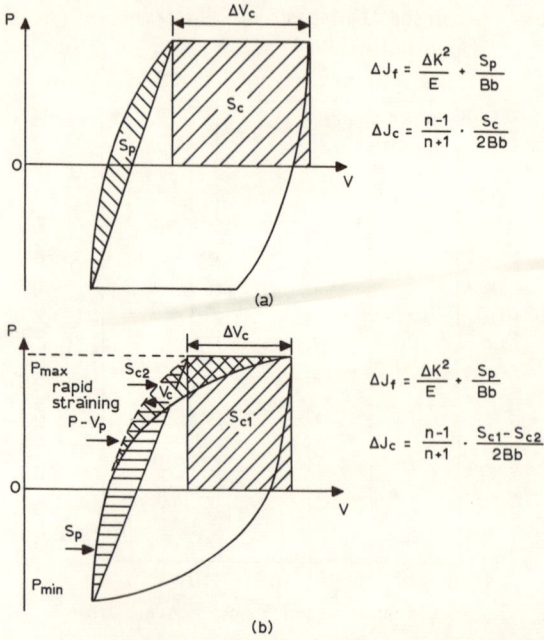

$$\Delta J_f = \frac{\Delta K^2}{E} + \frac{S_p}{Bb}$$

$$\Delta J_c = \frac{n-1}{n+1} \cdot \frac{S_c}{2Bb}$$

(a)

$$\Delta J_f = \frac{\Delta K^2}{E} + \frac{S_p}{Bb}$$

$$\Delta J_c = \frac{n-1}{n+1} \cdot \frac{S_{c1} - S_{c2}}{2Bb}$$

(b)

FIG. 11. Evaluation method for $\Delta J_f$ and $\Delta J_c$ on the basis of the load and crack center opening displacement: (a) trapezoidal stress waveform; (b) general stress waveform.

blunting and resharpening of the crack tip in each cycle, were observed on the fracture surface. It should be noted that the crack propagation was cycle-dependent and completely different from the creep (time-dependent) type. The confusion between these types has often created the misunderstandings of high temperature creep–fatigue strength. In this chapter, the cycle-dependent and the time-dependent phenomena are strictly distinguished from each other and are called fatigue and creep (in a wide sense), respectively. The fast crack propagation rate in the higher stress condition for an equal $\Delta K$ or $\Delta K_{eff}$ is due to the large-scale plasticity. Therefore, this should belong to the category of low cycle fatigue.

## CRACK PROPAGATION UNDER LOW CYCLE FATIGUE

The fatigue $J$-integral range (cyclic $J$-integral) $\Delta J_f$ was proposed by Dowling as a controlling parameter for $dl/dN$ not only in high cycle fatigue (elastic fatigue) but also in low cycle fatigue (elastic–plastic fatigue) at room temperature [33, 34]. Figure 14 shows the relationship between $dl/dN$ and $\Delta J_f$ of the tests shown in Fig. 13. The fatigue $J$-integral range is also valid in high temperature fatigue. The relationship is formulated as

$$dl/dN = C_f \Delta J_f^{m_f} \qquad (7)$$

The validity of $\Delta J_f$, however, becomes questionable under the lower frequency and/or higher mean stress condition, because the creep effect is marked. The interaction between creep and fatigue becomes the most important problem under such conditions, and is discussed in the following section.

Figure 15 shows the $dl/dN$–$\Delta J_f$ relationship for metals in the cycle-dependent fatigue condition obtained by the authors and their co-workers [14].

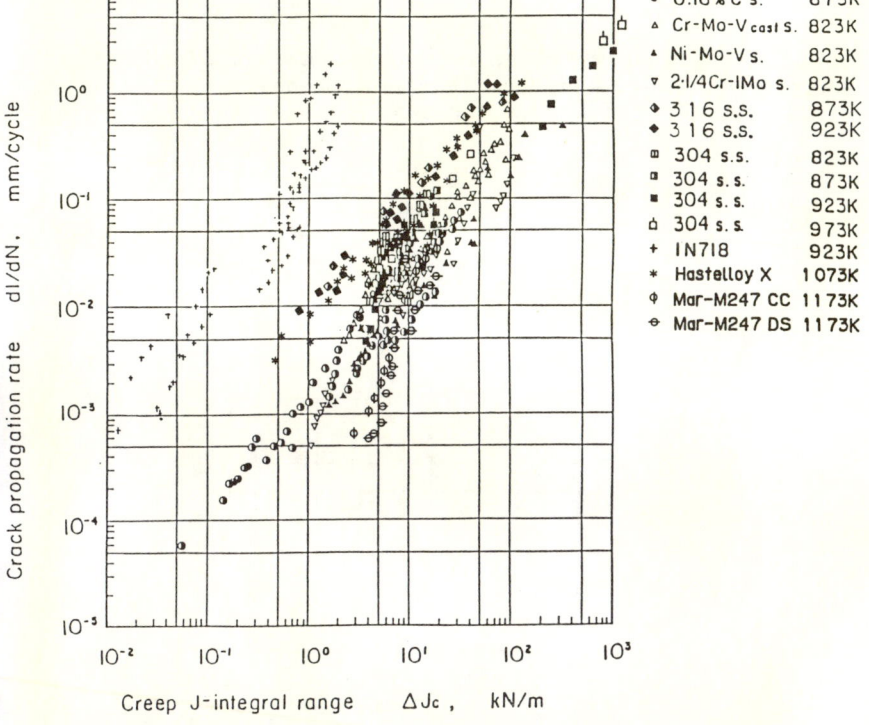

FIG. 12. Relationship between $dl/dN$ and $\Delta J_c$ in creep with a cyclic stress change.

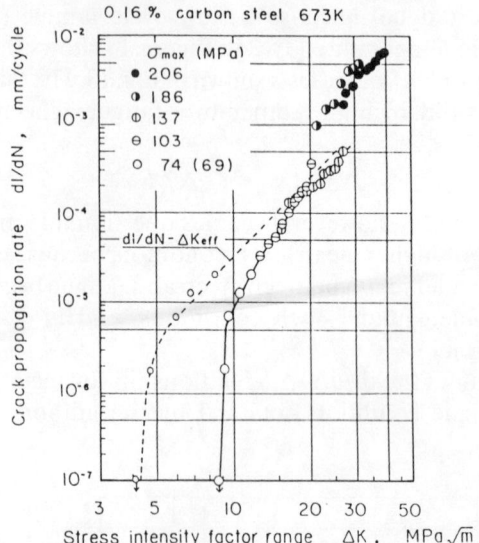

FIG. 13. Relationship between the crack propagation rate d$l$/d$N$ and the stress intensity factor range $\Delta K$ in fatigue. The dashed line shows d$l$/d$N$ as a function of the effective stress intensity factor range $\Delta K_{\text{eff}}$.

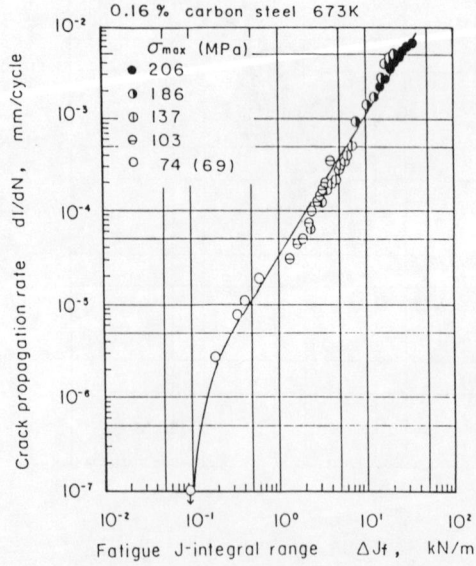

FIG. 14. Relationship between the crack propagation rate d$l$/d$N$ and the fatigue $J$-integral range $\Delta J_{\text{f}}$ in the fatigue of 0·16% carbon steel.

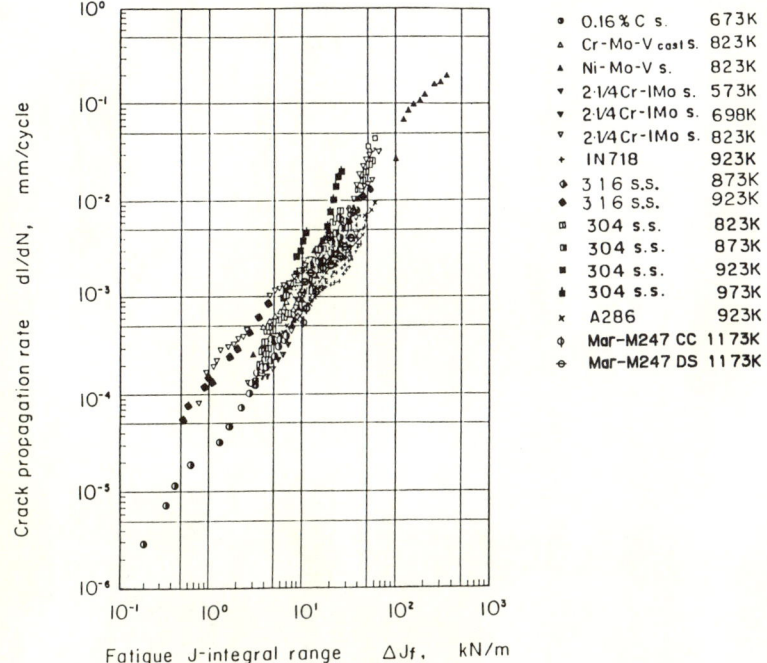

FIG. 15. Relationship between d$l$/d$N$ and $\Delta J_f$ in the fatigue of metallic materials.

## CRACK PROPAGATION UNDER THERMAL FATIGUE

The crack propagation rate d$l$/d$t$ of creep shows a strong dependence on the temperature, and its activation energy is almost equal to that of creep deformation [16]. The creep $J$-integral $J^*$, which is directly related to the crack tip creep rate, has also almost the same activation energy [16]. As a result, the d$l$/d$t$-$J^*$ relationship hardly depends on temperature. In addition, d$l$/d$N$ as well as $\Delta J_f$ are almost independent of temperature in fatigue. The independence of the d$l$/d$t$-$J^*$ and the d$l$/d$N$-$\Delta J_f$ relationships from the temperature suggests the applicability of the crack propagation laws (Eqns. (5) and (7)) to thermal fatigue.

In order to simulate the crack propagation of thermal fatigue, crack propagation tests were carried out using a tubular specimen of 0·16% carbon steel under the specific stress–temperature conditions shown in Fig. 16 [22]. The creep effect was stronger than the fatigue effect in tests I1, I2, O3 and O4, and vice versa in tests O1 and O2, which can be judged

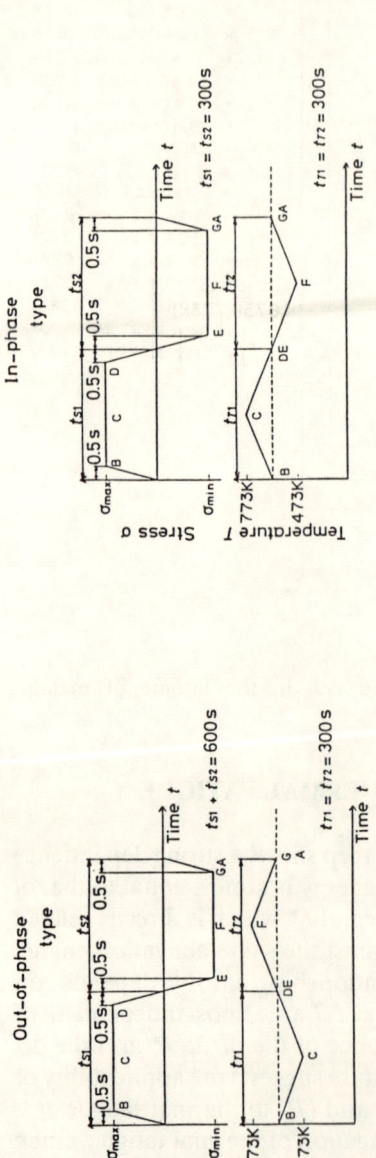

| Test no. | Type | Maximum stress (MPa) | Stress ratio | $t_{s1}$ (s) | $t_{s2}$ (s) | $T_{up} = T_{down}$ (K) | Symbol |
|----------|------|----------------------|--------------|--------------|--------------|-------------------------|--------|
| O1 | Out of phase | 211 | −0·58 | 100 | 500 | 523 | ◄ |
| O2 | Out of phase | 211 | −0·58 | 200 | 400 | 573 | ◄ |
| O3 | Out of phase | 206 | −0·62 | 300 | 300 | 623 | ◁ |
| O4 | Out of phase | 206 | −0·62 | 500 | 100 | 723 | ◁ |
| I1 | In phase | 147 | −1 | 300 | 300 | 623 | ▷ |
| I2 | In phase | 176 | −1 | 300 | 300 | 623 | ▽ |

$T_{max} = 773\,K$; $T_{min} = 473\,K$; $t_{T1} = t_{T2} = 300\,s$.

FIG. 16. Stress and temperature waveforms for thermal fatigue simulation tests.

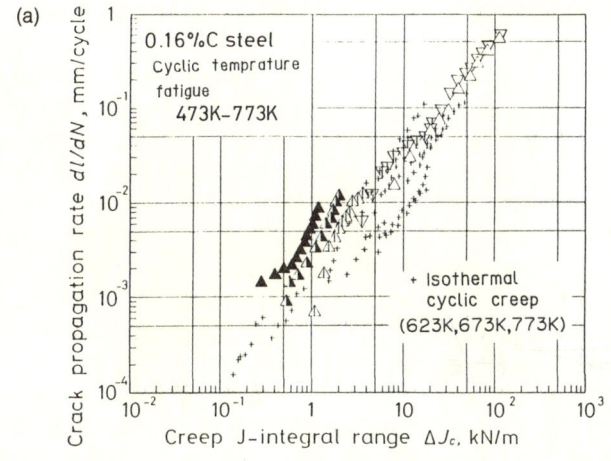

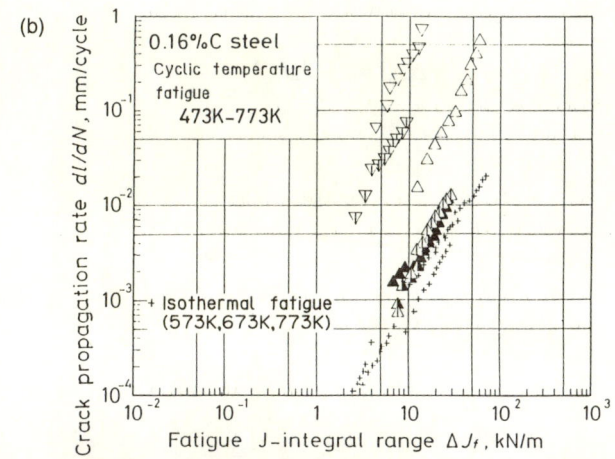

FIG. 17. Relationships between the crack propagation rate and $J$-integrals (0·16% C steel cyclic temperature fatigue 473–773 K): (a) $dl/dN$ versus $\Delta J_c$ relationship; (b) $dl/dN$ versus $\Delta J_f$ relationship.

from the boundary equation between creep and fatigue (Eqn. (8)), described in the next section. Figure 17 shows the $dl/dN–\Delta J_c$ relationship and the $dl/dN–\Delta J_f$ relationship, the symbol + representing the results of isothermal cyclic creep and fatigue. $dl/dN$ correlates with $\Delta J_c$ in the former group and with $\Delta J_f$ in the latter. Moreover, it signifies that the crack propagation of thermal fatigue obeys the same fracture mechanics laws as that of isothermal creep and fatigue.

## CRACK PROPAGATION LAWS AT HIGH TEMPERATURE

Crack propagation at high temperature is mainly classified into two types which are time dependent (creep) and cycle dependent (fatigue). The crack propagation laws are formulated as Eqns. (5) and (7). In order to clarify the interaction and the boundary between creep and fatigue, crack propagation tests on a $1Cr$-$1Mo$-$\frac{1}{4}V$ steel were carried out at 823 K (550 °C) using a trapezoidal stress waveform (Fig. 2(b)) with various stress hold times from 1 s to 1 h, and a sinusoidal stress waveform with a frequency of 1 Hz [19]. The crack propagation rate $dl/dN$ was plotted against $\Delta J_c$ under the constant $\Delta J_f$ (= 10 kN m$^{-1}$) condition shown in Fig. 18. This expression gives $dl/dN$ as constant for $\Delta J_c < 1$ kN m$^{-1}$ in fatigue and as proportional to $\Delta J_c$ for $\Delta J_c > 1$ kN m$^{-1}$ in creep. It is obvious from Fig. 18 that the behavior is clearly divided into the two regions of creep and fatigue. The boundary is formulated from Eqns. (5) and (7) as [19]

$$\Delta J_c = \frac{C_f}{C_c} \Delta J_f^{m_f} \qquad (8)$$

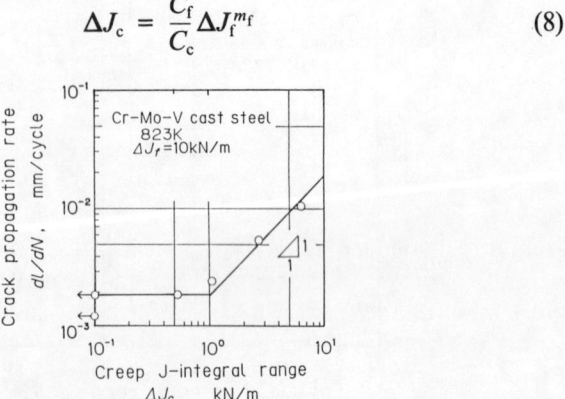

FIG. 18. Relationship between the crack propagation rate and the creep $J$-integral range under a constant fatigue $J$-integral range.

It is noteworthy that no or little mixed region exists between fatigue and creep. In other words, the so-called creep–fatigue interaction has no or little influence on the crack propagation laws (Eqns. (5) and (7)). It does not mean, however, that the $dl/dN$–$1/\nu$ (where $\nu$ is frequency) relationship for a given crack length is clearly divided into two regions where $dl/dN$ is constant and proportional to $1/\nu$, because the values of $\Delta J_f$ and $\Delta J_c$ are strongly affected by the interaction between the creep deformation and elastic–plastic deformation. In fact, an intermediate

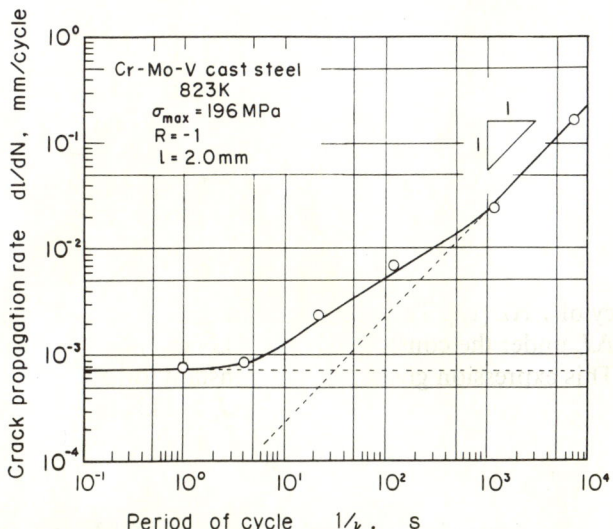

FIG. 19. Relationship between the crack propagation rate d$l$/d$N$ and the cyclic period 1/$v$.

region appears at the intermediate frequency of those tests as shown in Fig. 19 [19]. One of the typical interactions between them is the acceleration of creep rate caused by dynamic recovery during the previous compression period, the compression deformation markedly changing the creep constitutive equation of the following tension period. For example, the effect appears as the stress ratio effect of cyclic creep on d$l$/d$t$ and $\dot{\varepsilon}_{ct}$ shown in Figs. 8 and 9. Again, attention should be paid to the fact that it has no effect on the d$l$/d$t$–$J^*$ relationship shown in Fig. 10. The meaning of the difference between Figs. 18 and 19 can be similarly understood.

It is concluded that the deformation resistance (or the values of the two kinds of $J$-integrals) is strongly affected by creep–fatigue interaction, but the effect on the crack propagation resistance (or the fracture mechanics laws) is quite small. The characterization of high temperature strength is schematically summarized in Fig. 20.

## LIFE LAWS FOR A SMOOTH SPECIMEN

At high temperatures, some small cracks often initiate on the surface of a smooth specimen at a very early stage of the fatigue life and bring

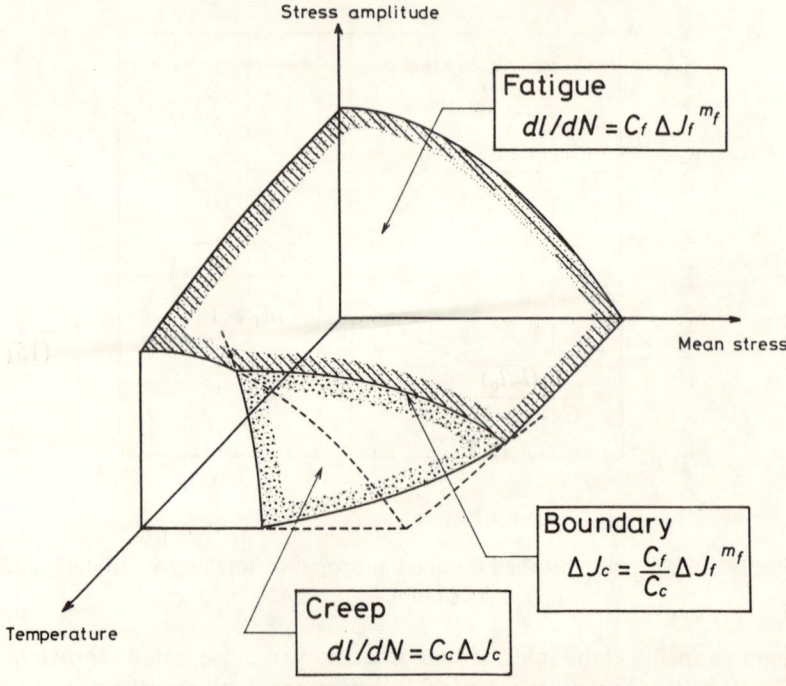

FIG. 20. Characterization of high temperature strength on the basis of the crack propagation laws.

about failure [35, 36]. In such a case, the life laws for the smooth specimen can be derived from the crack propagation laws described in the previous section. Carrying through the integral of Eqns. (5) and (7) from the initial crack length $l_0$ to the final crack length $l_f$ the following laws are obtained [14, 15 20]:

$$\Delta \tilde{W}_c N_f = D_c \qquad \text{for creep} \qquad (9)$$

$$\Delta \tilde{W}_f^{m_f} N_f = D_f \qquad \text{for fatigue} \qquad (10)$$

where $N_f$ is the failure and $\Delta \tilde{W}_c$ and $\Delta \tilde{W}_f$ are the strain energy parameters calculated by the following equations:

$$\Delta \tilde{W}_c = \frac{\sigma_{\max}}{\Delta \sigma} \frac{\sigma_{\max} \Delta \varepsilon_e}{2} + \frac{n+2}{n+3} \frac{n+1}{2\pi} f(n) \frac{\sigma_{\max} \Delta \varepsilon_c}{n+1} \qquad (11)$$

$$\Delta \tilde{W}_f = \frac{\Delta \sigma \Delta \varepsilon_e}{2} + \frac{n'+1}{2\pi} f(n') \frac{\Delta \sigma \Delta \varepsilon_p}{n'+1} \qquad (12)$$

$$f(n) = 3 \cdot 85 \sqrt{n}(1 - 1/n) + \pi/n \tag{13}$$

Here, $\sigma_{max}$ is the maximum stress, $\Delta\sigma$ is the stress range, $n$ and $n'$ are the stress exponents of power law creep and power law plasticity, $\Delta\varepsilon_e$, $\Delta\varepsilon_p$ and $\Delta\varepsilon_c$ are the elastic, plastic and creep strain ranges, respectively, and

$$D_c = \frac{\ln(l_f/l_0)}{C_c \times 2\pi M_J} \tag{14}$$

$$D_f = \frac{l_0^{1-m_f} - l_f^{1-m_f}}{C_f(m_f - 1)(2M_J)^{m_f}} \qquad m_f \neq 1$$

$$D_f = \frac{\ln(l_f/l_0)}{C_f \times 2\pi M_J} \qquad m_f = 1 \tag{15}$$

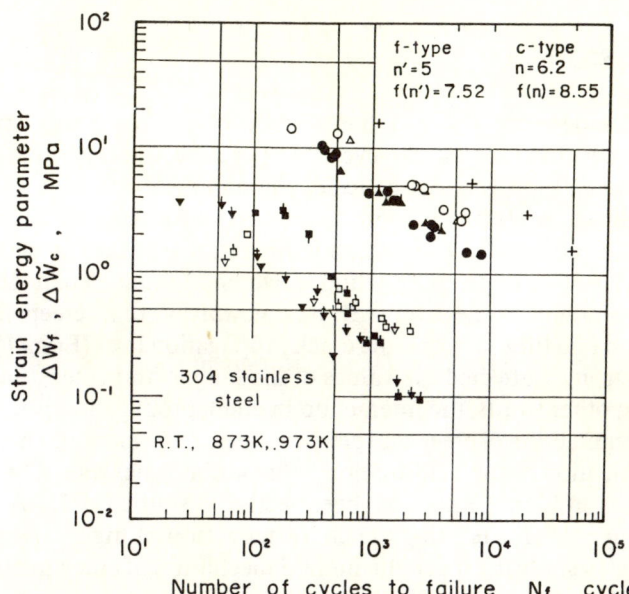

FIG. 21.  Relationships between the strain energy parameters and the number of cycles to failure for a smooth specimen.

where $M_J$ is the crack-shape and boundary correction factor. Details of the derivation and the general formulae (including the case of $m_c \neq 1$ in Eqn. (2)) have been reported in the previous paper [15]. The boundary between creep and fatigue is given as [18]

$$\Delta \tilde{W}_c = \frac{D_c}{D_f} \Delta \tilde{W}_f^{m_f} \qquad (16)$$

Figure 21 shows the experimental results for a type 304 stainless steel at room temperature, 873 K (660 °C) and 973 K (700 °C) in creep-fatigue tests, using asymmetric and symmetric triangular strain waveforms with various strain rates. Two types of relationships, which are Eqns. (9) (the creep type) and (10) (the fatigue type), can obviously be obtained from Fig. 21. The authors and their co-workers have verified the life laws for 13 kinds of steels and alloys, including low alloy steels, stainless steels and superalloys [14]. Moreover, the life laws have been also confirmed in thermal fatigue for several steels and alloys [14]. This suggests the applicability of characterizing, on the basis of crack propagation laws, by the life laws of a smooth specimen at high temperatures.

## CONCLUSION

High temperature strength has been divided into many specific regions according to the stress-cycle, temperature and time conditions. In this chapter they were synthetically characterized on the basis of the crack propagation behavior, with the results illustrated in Fig. 20. The crack propagation behavior at high temperature was mainly classified into the two types of creep and fatigue, which are time dependent and cycle dependent, respectively. It is noteworthy that creep-fatigue interaction had little effect on the crack propagation laws (Eqns. (5) and (7)), but strongly affected the values of the fatigue and creep $J$-integral ranges. In other words, the interaction in crack propagation is no more than the interaction of creep and elastic-plastic deformation. Therefore, characterization of the deformation behavior becomes one of the most important problems and has been actively studied [37]. Another interesting point for refining the characterization of high temperature strength is to study the mechanisms and mechanics of crack nucleation and small crack growth. Research on these is now under way and a systematic approach is expected in the near future.

# REFERENCES

[1]   S. Taira, R. Ohtani and T. Kitamura, *Trans. Am. Soc. Mech. Eng., Ser. H,* **101,** 154 (1979).
[2]   S. Taira, R. Ohtani and T. Komatsu, *Trans. Am. Soc. Mech. Eng., Ser. H,* **101,** 162 (1979).
[3]   S. Taira, R. Ohtani, T. Kitamura and K. Yamada, *J. Mater. Sci., Jpn.,* **28,** 414 (1979).
[4]   S. Taira, R. Ohtani, R. Shimizu, T. Kitamura and T. Kashiwagi, *Trans. Jpn. Soc. Mech. Eng.,* **46,** 468 (1980).
[5]   S. Taira, R. Ohtani, T. Yonekura, M. Osada and T. Kitamura, *Trans. Jpn. Soc. Mech. Eng.,* **46,** 861 (1980).
[6]   S. Taira, R. Ohtani and F. Wada, *Trans. Jpn. Soc. Mech. Eng.,* **46,** 1213 (1980).
[7]   R. Ohtani, Creep in structures, *Proc. 3rd IUTAM Symp., Leicester, UK,* Springer, Berlin, 542 (1981).
[8]   R. Ohtani, T. Kitamura and K. Yamada, *Proc. USA-Japan Joint Seminar, Hayama,* Nijhoff, 263 (1981).
[9]   R. Ohtani, K. Yamada, T. Kashiwagi and H. Matsubara, *Trans. Jpn. Soc. Mech. Eng.,* **A48,** 1378 (1982).
[10]  A. Nitta, T. Kitamura and K. Kuwabara, *J. Mater. Sci., Jpn.,* **33,** 185 (1984).
[11]  K. Kuwabara, A. Nitta, and T. Kitamura *J. Mater. Sci., Jpn.,* **33,** 338 (1984).
[12]  T. Kitamura, A. Nitta, and K. Kuwabara *J. Mater. Sci., Jpn.,* **33,** 585 (1984).
[13]  K. Kuwabara, A. Nitta, T. Kitamura and T. Ogata, *ASTM STP 924,* presented at *Int. Symp. on Fundamental Questions and Critical Experiments on Fatigue, Dallas-Fort Worth, TX, 1984,* to be published.
[14]  R. Ohtani, T. Kitamura, A. Nitta and K. Kuwabara, *ASTM STP 942,* presented at *Int. Symp. on Low Cycle Fatigue, Sagamore, New York, 1985,* to be published.
[15]  R. Ohtani and T. Kitamura, *J. Mater. Sci., Jpn.,* **34,** 843 (1985).
[16]  T. Kitamura, *Doctoral thesis,* Kyoto University, (1986).
[17]  T. Kitamura, H. Sugihara and R. Ohtani, *J. Mater. Sci., Jpn.,* **35,** 259 (1986).
[18]  R. Ohtani and T. Kitamura, *Proc. Int. Conf. on Role of Fracture Mechanics in Modern Technology, Fukuoka, Japan, 1986,* North-Holland, Amsterdam, 353 (1987).
[19]  T. Kitamura and R. Ohtani, *Trans. Jpn. Soc. Mech. Eng.,* **52A,** 1816 (1986).
[20]  R. Ohtani, T. Kitamura and T. Kinami, *J. Iron Steel Inst. Jpn.,* **72,** 711 (1986).
[21]  R. Ohtani, T. Kitamura and S. Enomoto, *Proc. 24th Symp. on High Temperature Strength, Japan,* Society of Materials Science, Kyoto, Japan, 56 (1986).
[22]  R. Ohtani, T. Kitamura and N. Tada, *ICM5, Beijing, China,* preprints, **2,** 1101 (1987).
[23]  K. Ohji, K. Ogura and S. Kubo, *Trans. Jpn. Soc. Mech. Eng.,* **42,** 350 (1976).
[24]  J. D. Landes and J. A. Begley, *ASTM STP 590,* 128 (1976).
[25]  K. M. Nikbin, G. A. Webster and C. E. Turner, *ASTM STP 601,* 47 (1976).
[26]  M. P. Harper and E. G. Ellison, *J. Strain Analysis,* **12,** 167 (1977).
[27]  K. Ohji, K. Ogura and S. Kubo, *Trans. Jpn. Soc. Mech. Eng.,* **44,** 1831, (1978).
[28]  R. Koterazawa, *J. Mater. Sci., Jpn.,* **8,** 209 (1959).

[29]  K. Ohji, K. Ogura and S. Kubo, *J. Mater. Sci., Jpn.,* **29**, 456 (1980).
[30]  H. Riedel and J. R. Rice, *ASTM STP 700,* 112 (1980).
[31]  G. R. Halford, M. H. Hirschberg and S. S. Manson, *ASTM STP 520,* 658 (1973).
[32]  S. S. Manson, *ASTM STP 520,* 744 (1973).
[33]  N. E. Dowling and J. A. Begley, *ASTM STP 590,* 82 (1976).
[34]  N. E. Dowling, *ASTM STP 601,* 19 (1976).
[35]  S. Usami, Y. Fukuda and S. Shida, *J. Mater. Sci., Jpn.,* **33**, 685 (1984).
[36]  R. Ohtani, T. Kinami and H. Sakamoto, *Trans. Jpn. Soc. Mech. Eng.,* **52A**, 1824, (1986).
[37]  T. Inoue, *J. Mater. Sci., Jpn.,* **32**, 594 (1983).

# Fracture Mechanics Evaluation of Crack Growth Behavior Under Creep and Creep–Fatigue Conditions

KIYOTSUGU OHJI and SHIRO KUBO

*Department of Mechanical Engineering, Osaka University, 2–1 Yamadaoka, Suita, Osaka, 565 Japan*

## ABSTRACT

The crack growth behaviour of creep-ductile type 304 stainless steels under creep and creep–fatigue conditions was investigated. It was found that the creep crack growth rates $da/dt$ were best correlated with the modified $J$-integral $J^*$ (or $C^*$-integral). The crack growth under creep–fatigue conditions was either purely time dependent or purely cycle dependent, depending on the test conditions employed. The time-dependent crack growth rate and cycle-dependent crack growth rate were expressed as a sole function of the modified $J$-integral $J^*$ and the effective $J$-integral range $\Delta J_f$, respectively, irrespective of the stress waveforms and loading conditions applied. The fracture surfaces were intergranular and transgranular for the time-dependent and the cycle-dependent crack growths, respectively. A threshold for crack growth was observed in the cycle-dependent crack growth under creep–fatigue conditions. In the sub-threshold region, time-dependent crack growth with 'pseudo-thresholds' appeared. The transition between the time-dependent and cycle-dependent crack growth modes was found to be predictable on the basis of the competitive damage hypothesis. Based on this finding, a crack growth mode map was proposed. Crack closure was observed under negative stress ratios $R = \sigma_{min}/\sigma_{max}$. The behaviour of crack closure was found to be very complicated, and sensitive to the frequency and loading history.

## INTRODUCTION

An increasing number of industrial plants and components are operated at high temperatures where creep deformation is significant. To ensure the reliability of these plants and components, and to determine their proper inspection intervals, a quantitative evaluation of

91

the crack growth behaviour under creep and creep–fatigue conditions is indispensable.

The idea of applying fracture mechanics concepts to creep crack growth problems was first proposed by Siverns and Price [1] in 1970. In their paper they suggested the use of the stress intensity factor as a correlating parameter for creep crack growth rates. Since then, many papers have been published on creep crack growth problems, and various other correlating parameters such as the modified $J$-integral (which is also referred to as the $C^*$-integral), net section stress, reference stress and others have been proposed. The history of research and debate on this subject may be found in the reviews [2–12] published so far. Through these, the majority of researchers seems to have approved a general trend that the creep crack growth rates of ductile materials can be characterized by the modified $J$-integral (or $C^*$-integral).

When compared with the crack growth under pure creep conditions, the crack growth under creep–fatigue conditions is far more difficult because of the involvement of many factors which affect the behavior. Since the fundamental behavior of crack growth under pure creep conditions was already understood, crack growth under creep–fatigue conditions has received increasing attention of engineers and scientists with regard to strength problems at high temperatures. Many researchers are challenging this problem and many publications on this subject have appeared recently, so that a considerable amount of information relating to this subject has been accumulated. Several reviews [13–19] have already been published, but the situation is still confused, although some clues to the problems are being obtained.

For more than ten years, the present authors have devoted their continuous research effort to high-temperature fracture mechanics and its application to the characterization of crack growth behavior under creep and creep–fatigue conditions. This paper gives an overview of their group's principal results obtained in these fields with type 304 stainless steels.

## FRACTURE MECHANICS PARAMETERS AND THEIR EVALUATION

As a fracture mechanics parameter governing the stress and strain-rate fields near the creep crack tip, the modified $J$-integral $J^*$ (or $C^*$-integral) was introduced [20–23].

The value of $J^*$ can be experimentally determined from the net section stress $\sigma_{net}$ and the load-point displacement rate $\dot{\Delta}$ [24, 25]. For deep center-cracked specimens, the experimental $J^*$ value $J^*_{ex}$ is approximately given by (see Appendix I)

$$J^*_{ex} = \sigma_{net}\dot{\Delta} \tag{1}$$

Equation (1) is valid for shallow center-cracked specimens only if the load-point displacement rate $\dot{\Delta}$ is replaced by a crack-center opening displacement rate [25].

The analytical value $J^*_{an}$ of the modified $J$-integral based on the assumption of the predominance of Norton-type steady-state creep may be obtained using the following equation proposed by Ohji et al. [26]:

$$1/J^*_{an} = 1/J^*_s + 1/J^*_d \tag{2}$$

where $J^*_s$ and $J^*_d$ indicate $J^*$ values evaluated for shallow and deep cracks, respectively, and can be approximated by the following equation for center-cracked specimens subjected to tension:

$$
\begin{aligned}
J^*_s &= \pi(\pi n/2)^{(n-1)/(2n-1)}a\sigma_{net}\dot{\varepsilon}_{net} \\
J^*_d &= [4\pi/(\pi^2 - 4)\sqrt{n}]\,b\sigma_{net}\dot{\varepsilon}_{net}
\end{aligned}
\tag{3}
$$

In Eqn. (3), $a$ and $b$ denote the half crack length and the ligament length of the specimen, respectively. The strain rate $\dot{\varepsilon}_{net}$ in Eqn. (3) is calculated from $\sigma_{net}$ using the Norton-type creep law, $n$ denoting the creep exponent. For sinusoidal stress waves, the net section stress $\sigma_{net}$ in Eqn. (3) may be replaced by the equivalent stress for dynamic creep [27].

As correlating parameters of fatigue crack growth rates, the stress intensity factor range $\Delta K$ and the $J$-integral range $\Delta J_f$ were introduced. The value of $\Delta J_f$ was evaluated in accordance with the method proposed by Dowling [28].

## CHARACTERIZATION OF CREEP CRACK GROWTH BEHAVIOR

To identify experimentally the most probable mechanical parameter which controls the creep crack growth rates of type 304 stainless steels, deep center-cracked (DCN) specimens, shallow center-cracked (SCN) specimens, circumferentially cracked round bar specimens and compact-tension (CT) specimens were adopted [29–35].

Figure 1 shows the correlation curves of creep crack growth rate $da/dt$ with stress intensity factor $K_I$, net section stress $\sigma_{net}$ and the modified $J$-integral $J_{ex}^*$ obtained with DCN and SCN proportional specimens and CT specimens of type 304 stainless steel at 650°C [34, 35].

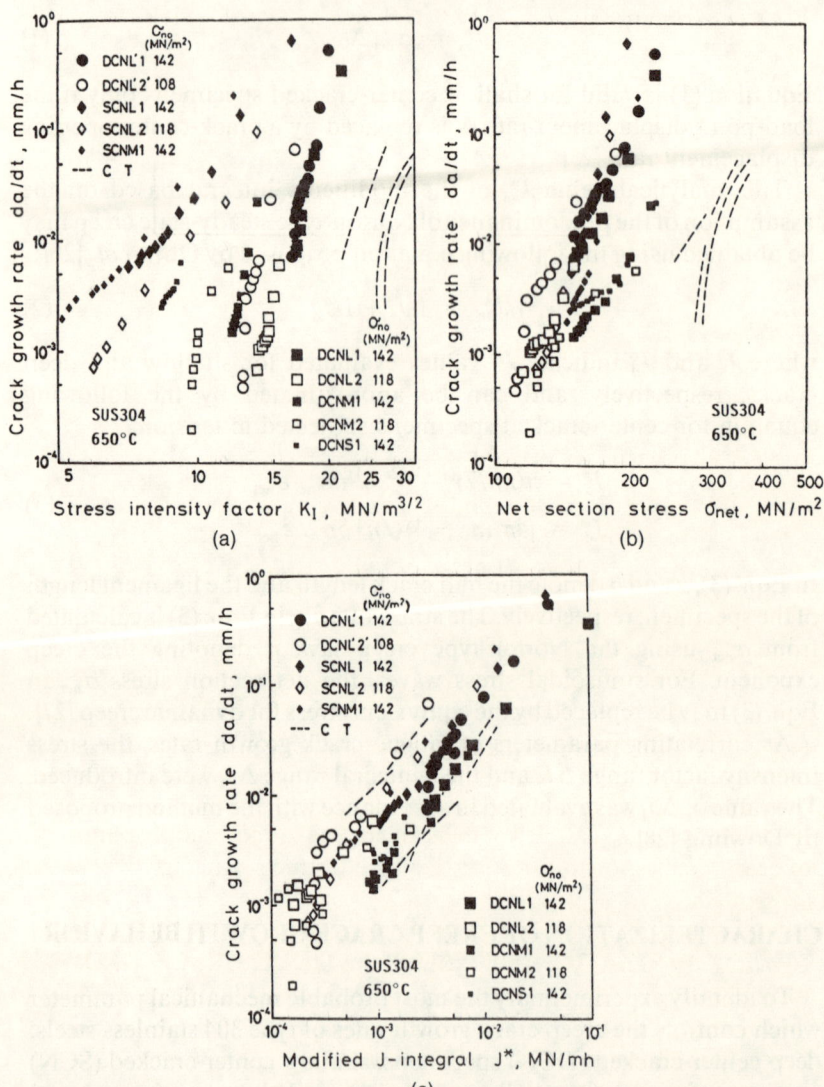

FIG. 1. Correlations of creep crack growth rates for type 304 stainless steel at 650°C with $K_I$, $\sigma_{net}$ and $J^*$: (a) $da/dt$-$K_I$; (b) $da/dt$-$\sigma_{net}$; (c) $da/dt$-$J^*$.

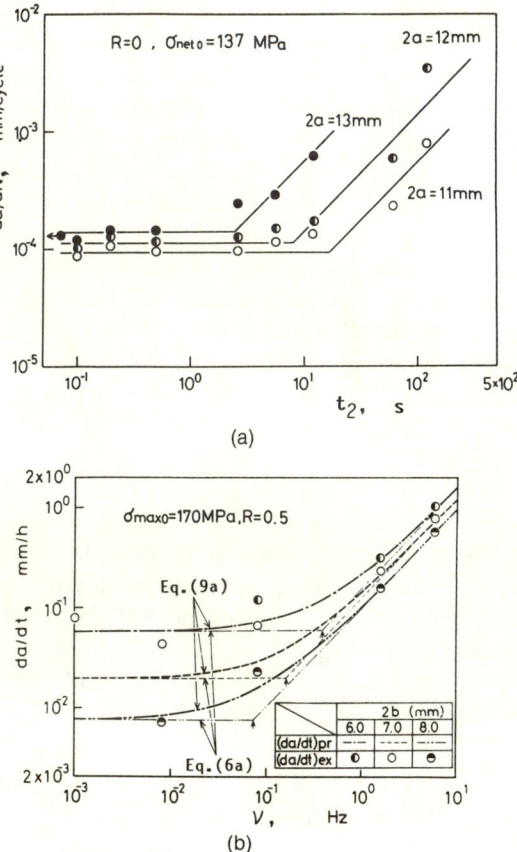

FIG. 2. Crack growth rates at different hold times or frequencies: (a) trapezoidal wave; (b) sinusoidal wave.

It is evident that $J^*$ is superior to $K_I$ and $\sigma_{net}$ as the correlating parameter of creep crack growth rates for this type of ductile material.

## CHARACTERIZATION OF CREEP-FATIGUE CRACK GROWTH BEHAVIOR

### Crack Growth Rates at Different Frequencies

A series of experiments were conducted at various frequencies under otherwise the same loading conditions. Figure 2(a) shows the crack growth rate per cycle $da/dN$ as a function of the hold time $t_2$ at several given crack lengths, obtained for a trapezoidal stress wave of zero stress ratio ($R = \sigma_{min}/\sigma_{max}$) [36]. The relationship between the crack growth rate in time $da/dt$ for a sinusoidal stress wave of positive $R$ is plotted against the cycle frequency $v$ in Fig. 2(b) [37, 38].

It can be seen from the figures that, when the hold time $t_2$ is small or the frequency $v$ is high, $da/dN$ is nearly constant, and $da/dt$ increases almost in proportion to $v$. In contrast, $da/dN$ is almost proportional to $t_2$ for large $t_2$ values, and $da/dt$ remains constant for low $v$ values.

These results indicate that the crack growth is purely cycle dependent at high frequencies, while at low frequencies it is purely time dependent. It can be seen from the figures that the width of the transition region between these two modes is smaller than two logarithmic cycles of $t_2$ or $v$, and that the deviation of $da/dN$ or $da/dt$ from the purely time-dependent or cycle-dependent crack growth rate is small.

### Characterization of Crack Growth Behavior in Terms of Elastic–Plastic Fracture Mechanics Parameters

Creep–fatigue crack growth experiments on center-cracked specimens made of type 304 stainless steels were conducted for many combinations of the loading stress level $\sigma_{max0}$, stress ratio $R$, frequency $v$ and stress waveform (sinusoidal or trapezoidal) [36–43]. The relationship between the crack growth rate $da/dt$ and $J_{ex}^*$ obtained under sinusoidal stress waves with non-negative stress ratios is illustrated in Fig. 3(a) [38, 41].

Under static creep conditions, $da/dt$ can be expressed solely as a function of $J_{ex}^*$ as [33–35]

$$(da/dt)_c = 8.0 \times 10^{-3} J_{ex}^* \qquad (4)$$

($(da/dt)_c$ in mm h$^{-1}$, $J_{ex}^*$ in kJ m$^{-2}$ h$^{-1}$) which is shown by a solid line in the figure. The two parallel broken lines in the figure indicate the limits of factors of 2 on $(da/dt)_c$.

It can be observed in Fig. 3(a) that data obtained under a high loading stress level $\sigma_{max0}$ and high $R$ values, and/or at low frequencies, including those which were concluded to be time-dependent from such $da/dt$–$v$ relationships as those shown in Fig. 2, fall within the scatter band for the static creep crack growth. These findings provide evidence for the fact that, in the time-dependent crack growth region, $da/dt$ can be solely expressed as a function of $J_{ex}^*$, irrespective of the loading waveform and values of $R$, $\sigma_{max0}$ and $v$.

Fractographical examinations showed that the fracture was of an intergranular (or creep) type for the time-dependent crack growth [40].

The crack growth rates in time $da/dt$ were transformed into cyclic

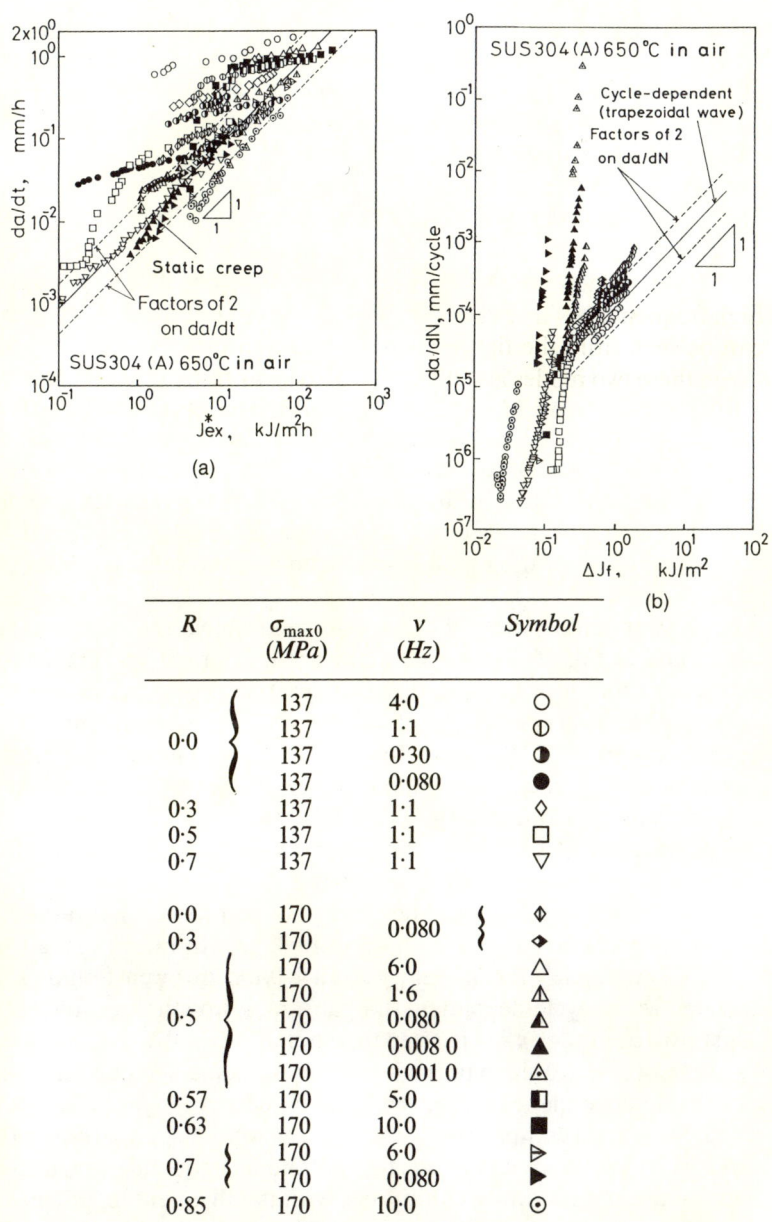

| R | $\sigma_{max0}$ (MPa) | $\nu$ (Hz) | | Symbol |
|---|---|---|---|---|
| 0·0 { | 137 | 4·0 | | O |
| | 137 | 1·1 | | ⊕ |
| | 137 | 0·30 | | ◑ |
| | 137 | 0·080 | | ● |
| 0·3 | 137 | 1·1 | | ◇ |
| 0·5 | 137 | 1·1 | | □ |
| 0·7 | 137 | 1·1 | | ▽ |
| 0·0 | 170 | 0·080 { | | ◈ |
| 0·3 | 170 | | | ◆ |
| 0·5 { | 170 | 6·0 | | △ |
| | 170 | 1·6 | | ◮ |
| | 170 | 0·080 | | ▲ |
| | 170 | 0·008 0 | | ▲ |
| | 170 | 0·001 0 | | ◭ |
| 0·57 | 170 | 5·0 | | ◧ |
| 0·63 | 170 | 10·0 | | ■ |
| 0·7 { | 170 | 6·0 | | ▷ |
| | 170 | 0·080 | | ▶ |
| 0·85 | 170 | 10·0 | | ⊙ |

FIG. 3. Crack growth rates under sinusoidal stress waves at 650 °C: (a) d$a$/d$t$–$J^*_{ex}$; (b) d$a$/d$N$–$\Delta J_f$.

crack growth rates $da/dN$ simply by using the relationship $da/dN = (da/dt)/v$. The values of $da/dN$ are plotted against $\Delta J_f$ in Fig. 3(b).

It has been reported that the relationship of the cycle-dependent crack growth rates $da/dN$ versus $\Delta J_f$ obtained under trapezoidal stress waves could be expressed as

$$(da/dN)_f = 1.2 \times 10^{-4} \Delta J_f \qquad \text{for } \Delta J_f > 0.1 \qquad (5)$$

($(da/dN)_f$ in mm cycle$^{-1}$, $\Delta J_f$ in kJ m$^{-2}$) which is shown by a solid line in Fig. 3(b). The range of factors of 2 on $(da/dt)_f$ is given by the dashed lines in the figure.

The values of $da/dN$ larger than $3 \times 10^{-5}$ mm cycle$^{-1}$ obtained under low stress and low $R$ values, and/or at high frequency $v$, fall between the broken lines. The $da/dN$ data, which were identified as cycle dependent from $da/dN$–$v$ relationships, also fall within this band. These observations lead to the conclusion that in the cycle-dependent crack growth region, the crack growth rate $da/dN$ can be expressed solely as a function of $\Delta J_f$, irrespective of the loading waveform and the values of $R$, $\sigma_{max0}$ and $v$.

The experimental points that were located above the creep crack growth band in Fig. 3(a) were found to be reduced to the points inside the data band for the cycle-dependent (fatigue) crack growth in Fig. 3(b). Similarly, the experimental points above the cycle-dependent crack growth band in Fig. 3(b) were found to be included in the time-dependent (creep) crack growth band in Fig. 3(a). Similar results have been obtained by Taira et al. [44] and Kitamura and Ohtani [45].

From fractographical observations, the fracture was transgranular for the cycle-dependent crack growth [40].

In the region where $\Delta J_f$ is smaller than about 0.1 kJ m$^{-2}$, a threshold-like behavior was observed [37,38,41]. Crack growth rates as small as $10^{-7}$ to $10^{-6}$ mm cycle$^{-1}$ for $\Delta J_f$ nearly equal to 0.1 kJ m$^{-2}$ were found to be associated with a cycle-dependent (fatigue) crack growth threshold [41].

As shown in Fig. 3(b), a very steep threshold-like $da/dN$–$\Delta J_f$ relationship can be observed even in the region where $\Delta J_f$ is smaller than the value for the cycle-dependent crack growth threshold $((\Delta J_f)_{th} = 0.1$ kJ m$^{-2}$).

Figure 4 shows a comparison of the crack growth rates $da/dt$ obtained in this sub-threshold region at two different frequencies $v$ under otherwise the same loading conditions, with the ligament length as an abscissa. The crack growth rates $da/dt$ are almost the same for the entire range of ligament length $2b$. These date fell within the data band of creep-type crack growth in the $da/dt$–$J_{ex}^*$ correlation (Fig. 3(a)).

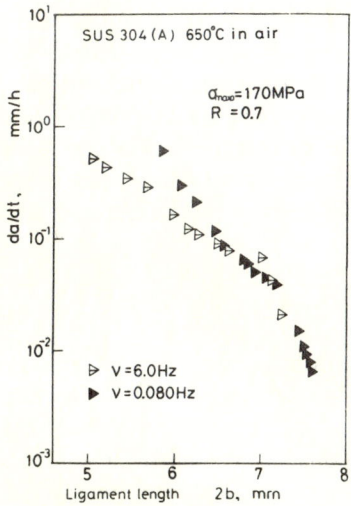

FIG. 4. Comparison of crack growth rates at different frequencies in the sub-threshold region.

Furthermore, intergranular fracture was predominant in this sub-threshold region, as shown in Fig. 5 [40].

These observations indicate that time-dependent crack growth emerged below the threshold level $(\Delta J_f)_{th}$ because the cycle-dependent crack growth was extinguished. This sub-threshold time-dependent crack growth yielded pseudo-thresholds as in Fig. 3(b), the values of the

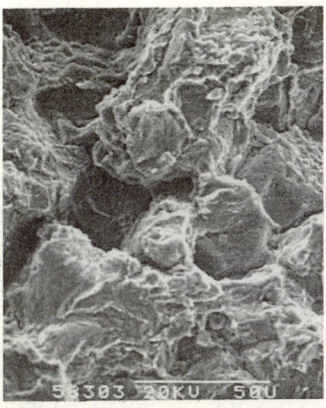

FIG. 5. Fracture surface observed in the time-dependent region appearing in the sub-threshold region of cycle-dependent crack growth.

pseudo-threshold being smaller than $(\Delta J_f)_{th}$ and arbitrary depending on the loading conditions [41]. This unusual threshold behavior may be called a creep-induced floating fatigue threshold [38].

These observations were made at 650 °C. The behavior of crack growth with type 304 stainless steel under creep–fatigue conditions was found to be essentially the same at temperatures ranging from 550 to 700 °C [43]. Figure 6 shows the $da/dt$–$J^*_{ex}$ and $da/dN$–$\Delta J_f$ correlations

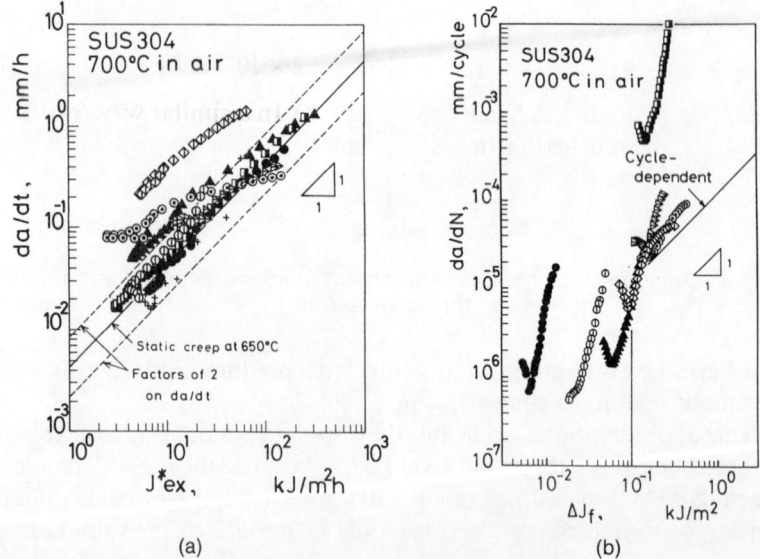

(a)                                          (b)

FIG. 6. Crack growth rates under creep–fatigue conditions at 700 °C: (a) $da/dt$–$J^*_{ex}$; (b) $da/dN$–$\Delta J_f$.

obtained at 700 °C. By comparing these figures with Fig. 3, it can be concluded that both the relationship between the purely time-dependent crack growth rate and $J^*_{ex}$ and the relationship between the purely cycle-dependent crack growth rate and $\Delta J_f$ are rather insensitive to temperature.

## Conditions for the Transition Between Cycle-Dependent and Time-Dependent Crack Growth Modes

The foregoing experimental observations suggest that the transition between the cycle-dependent and time-dependent crack growth modes may be determined by the competitive (or dominant) damage hypothesis that the crack growth mode, either time dependent or cycle dependent, which gives the higher crack growth rate dominates the whole phenomenon under creep–fatigue conditions. Then

$$da/dt = \max[(da/dt)_c, (da/dt)_f]  \qquad (6a)$$

or equivalently

$$da/dN = \max[(da/dN)_c, (da/dN)_f]  \qquad (6b)$$

In Eqn. (6a), $(da/dt)_f$ denotes the cycle-dependent crack growth rate in time and is given by converting $(da/dN)_f$ in Eqn. (5) into $(da/dt)_f$ as

$$(da/dt)_f = (da/dN)_f v = 1.2 \times 10^{-4} \, \Delta J_f v  \qquad (7)$$

($(da/dt)_f$ in mm h$^{-1}$, $\Delta J_f$ in kJ m$^{-2}$, $v$ in h$^{-1}$). In a similar way, $(da/dN)_c$ in Eqn. (6b) indicates the time-dependent cyclic crack growth rate and is calculated from $(da/dt)_c$ in Eqn. (4) as

$$(da/dN)_c = (da/dt)_c/v = 8.0 \times 10^{-3} J^*/v = 8.0 \times 10^{-3} \, \Delta J_c  \qquad (8)$$

($(da/dN)_c$ in mm cycle$^{-1}$, $J^*$ in kJ m$^{-2}$ h$^{-1}$, $v$ in h$^{-1}$, $\Delta J_c$ in kJ m$^{-2}$) where $\Delta J_c$, which is defined by $J^*/v$, denotes the time integral of $J^*$ over one stress cycle [44].

Another simple scheme for estimating $da/dt$ or $da/dN$ may be the summation damage hypothesis:

$$da/dt = (da/dt)_c + (da/dt)_f  \qquad (9a)$$

or

$$da/dN = (da/dN)_c + (da/dN)_f  \qquad (9b)$$

The summation damage hypothesis is almost equivalent to the competitive damage hypothesis, if factors of 2 are allowed for the experimental scatter of crack growth data.

The thin and bold lines in Fig. 2(b) indicate the values predicted by Eqns. (6a) and (9a), respectively. Agreement between the predicted and observed values of $da/dt$ is good, although the analytical $J^*$ value $J^*_{an}$ was used as $J^*_{ex}$ in Eqn. (4) to evaluate $(da/dt)_c$, instead of using the experimentally determined value $J^*_{ex}$.

By using the summation damage hypothesis, $da/dN$ can be readily predicted:

$$da/dN = (da/dN)_c + (da/dN)_f$$
$$= 8.0 \times 10^{-3}(\Delta J_c + 0.015 \, \Delta J_f)  \qquad (10)$$

($da/dN$ in mm cycle$^{-1}$, $\Delta J_c$ in kJ m$^{-2}$, $\Delta J_f$ in kJ m$^{-2}$).

Figure 7 shows a correlation of $da/dN$ with $(\Delta J_c + 0.015 \Delta J_f)$ for the sinusoidal stress waves. In the figure, $\Delta J_f$ was taken to be zero in the region of $\Delta J_f < 0.1$ kJ m$^{-2}$ in order to take into account the effect of the threshold in the cycle-dependent crack growth. The data points which were associated with the near-threshold cycle-dependent crack growth were omitted from the figure for simplicity. A good correlation can be observed.

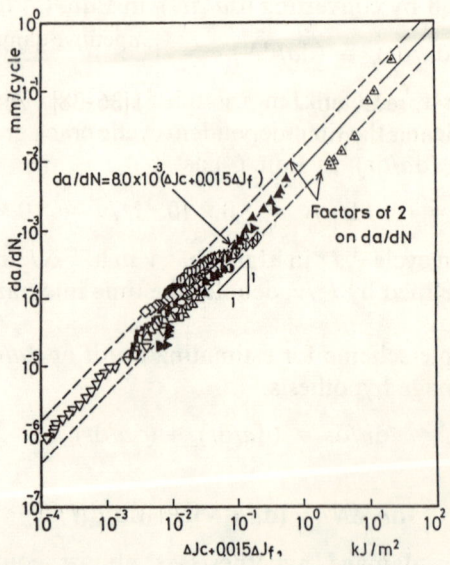

FIG. 7. Prediction of fatigue crack growth rates under creep–fatigue conditions.

Figure 8 shows schematically the crack growth mode map [41] constructed on the basis of the competitive damage hypothesis (Eqn. (6)). The transition condition between the time-dependent and cycle-dependent crack growth modes is determined by equating Eqns. (5) and (8), and is expressed as

$$
\begin{aligned}
J^*/v\,\Delta J_f &= \Delta J_c/\Delta J_f = 0.015, & \Delta J_f &> 0.1 \\
J^*/v\,\Delta J_f &= \Delta J_c/\Delta J_f < 0.015, & \Delta J_f &= 0.1
\end{aligned}
\tag{11}
$$

The effect of the stress level, crack length and $R$ on the transition between the time-dependent and cycle-dependent crack growth modes

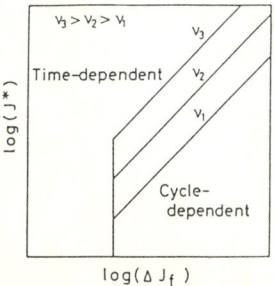

FIG. 8. Crack growth mode map based on the competitive damage hypothesis.

was successfully predicted by Eqn. (11) [36–38]. As an example, separation of the crack growth modes based on this map was made by using the results shown in Fig. 3 [37,38]. Experimental data which satisfied $\Delta J_c/\Delta J_f > 0.015$ or $\Delta J_f < 0.1$ (the time-dependent mode) were plotted against $J_{ex}^*$ in Fig. 9(a), while those which satisfied $\Delta J_c/\Delta J_f < 0.015$ and $\Delta J_f > 0.1$ (the cycle-dependent mode) were plotted against $\Delta J_f$ in Fig. 9(b). These figures provide evidence for the validity of the transition condition and the crack growth mode map proposed by the present authors' group.

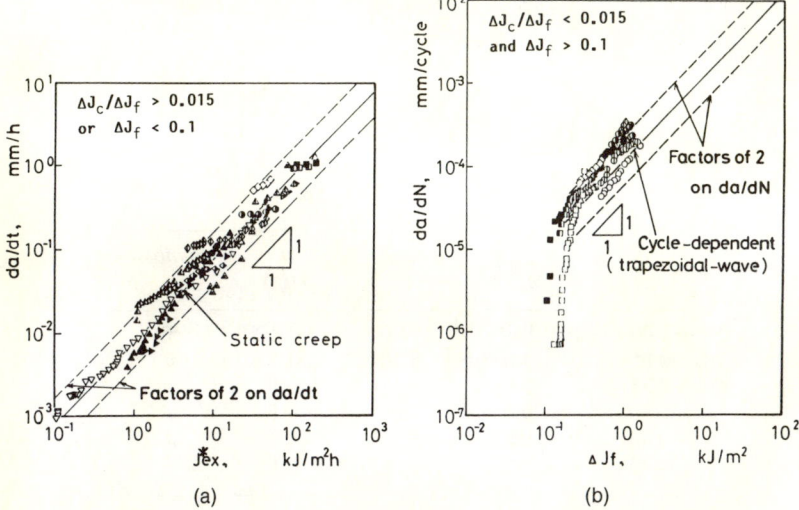

FIG. 9. Separation of the cycle-dependent and time-dependent crack growth modes: (a) $\Delta J_c/\Delta J_f > 0\cdot015$ or $\Delta J_f < 0\cdot1$; (b) $\Delta J_c/\Delta J_f < 0\cdot015$ and $\Delta J_f > 0\cdot1$.

## Crack Closure under Creep–Fatigue Conditions

It was found that crack closure had a serious effect on the crack growth under creep–fatigue conditions [42]. No crack closure was observed at high values of $R \geqslant 0.3$–$0.5$, while crack closure was observed at negative values of $R$. Figures 10 and 11 show the behavior of the crack growth rate $da/dN$ and the corresponding behavior of the effective stress range ratio ($U = \Delta\sigma_{eff}/\Delta\sigma$), respectively, with the number of cycles $N$, obtained for type 304 stainless steel at 650 °C [42].

In the test with $\nu = 0.1$ Hz, the $U$ values for $R \geqslant -0.75$ decreased with an increase in $N$, and reached very low particular stabilized values. On the other hand, the $U$ values for $R = -0.85$ at $\nu = 0.1$ Hz and also for $R = -1.0$ at $\nu = 0.2$ Hz changed from decreasing to increasing at small $N$ and large $U$ values, and reached unity.

By comparing Figs. 10 and 11, it was concluded that the behavior of $da/dN$ resembles that of $U$.

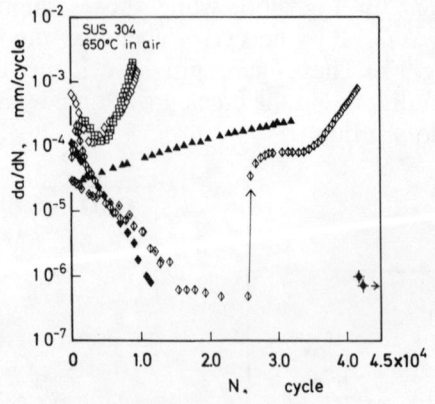

| $R$ | $\sigma_{max0}$ (MPa) | $\sigma_{a0}$ (MPa) | $\nu$ (Hz) | Symbol |
|---|---|---|---|---|
| −1·00 | 94 | 94·0 | 0·2 | ⊞ |
| −0·85 | 94 | 87·0 | 0·1 | ◇ |
| −0·75 | 94 | 82·3 | 0·1 | ◈ |
| −0·50 | 94 | 70·5 | 0·1 | ◈ |
| −0·40 | 94 | 65·8 | 1·0 | ▲ |
| −0·40 | 94 | 65·8 | 0·1 | ◆ |

FIG. 10. Behavior of $da/dN$ associated with crack closure under creep–fatigue conditions.

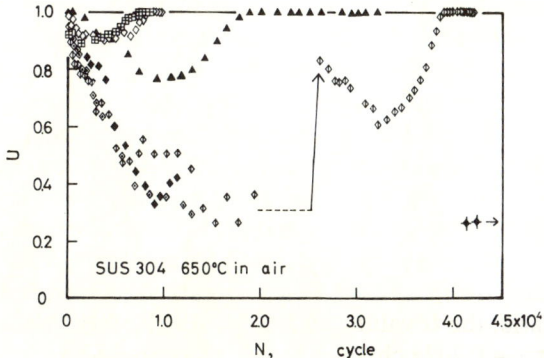

FIG. 11. Crack closure behavior under creep–fatigue conditions.

In the two tests at $R = -0.40$, the $U$ values at $v = 1.0$ Hz were much larger than those at $v = 0.1$ Hz. The former changed to increasing at relatively small $N$, while the latter kept decreasing to a low stabilized value. In the latter case, $da/dN$ decreased to values as low as $10^{-6}$ mm cycle$^{-1}$ and the crack almost ceased to propogate.

When the test at $R = -0.75$ was interrupted for 42 h at $N = 26\,000$ cycles, with $da/dN < 10^{-6}$ mm cycle$^{-1}$ and keeping the applied stress zero and the temperature at 650 °C, the $U$ value recovered considerably and $da/dN$ jumped to a value two orders of magnitude higher than that before the interruption.

These anomalous effects of frequency and loading history were attributed to the interaction between crack growth and oxide-induced crack closure enhanced by fretting [42].

It was found that the relationship between the cycle-dependent crack growth rate $da/dN$ and stress intensity factor range $\Delta K$ obtained for $-1 \leqslant R < 0.2$ deviated from the $da/dN$–$\Delta K$ correlation obtained under $0.5 \leqslant R$ where crack closure was not expected. The values for the fatigue threshold $\Delta K_{th}$ for low $R$ were much greater than those obtained for high $R$. Such a strong $R$ dependence of the crack growth behavior was eliminated by using the $J$-integral range $(\Delta J_f)_{eff}$, in which the effect of crack closure and plasticity were taken into account.

### Acceleration of $J^*$

It has been shown that the $da/dt$–$J_{ex}^*$ relationship was insensitive to temperatures and loading conditions such as frequency and waveform.

However, this does not necessarily mean that $da/dt$ and $J_{ex}^*$ are insensitive to temperatures, frequencies and waveforms. These factors may accelerate the value of $J_{ex}^*$ and hence $da/dt$, even when a unique correlation is observed between $da/dt$ and $J_{ex}^*$.

Figure 12(a) shows a comparison of $J_{ex}^*$ with $J_{an}^*$ obtained at 650°C with type 304 stainless steel [43]. It was found that the value of $J_{ex}^*$ was much higher than that of $J_{an}^*$ when the frequency $v$ was high and the crack growth mode was cycle dependent, although the high values of $J_{ex}^*$ under these conditions did not lead to the acceleration of $da/dt$ because of the predominance of the cycle-dependent crack growth mode. On the other hand, $J_{ex}^*$ was nearly equal to $J_{an}^*$ when the time-dependent crack growth prevailed.

Figure 12(b) shows the same kind of relationship obtained at 600°C [43]. It can be seen that $J_{ex}^*$ is much higher than $J_{an}^*$, which was evaluated based on the steady-state creep assumption, even when the crack growth was purely time-dependent. Dynamic strain aging may be responsible for this increase in $J_{ex}^*$.

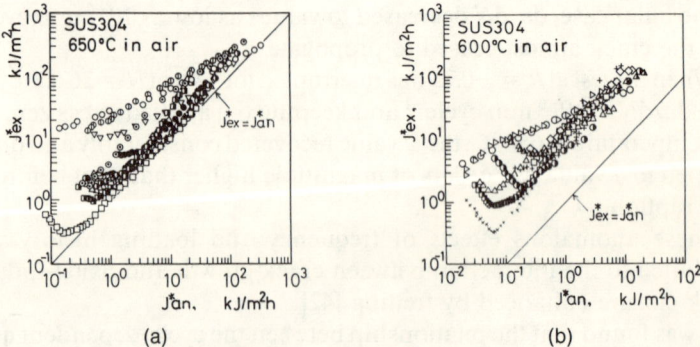

FIG. 12. Comparison of the experimental $J^*$ values $J^*_{ex}$ and the analytical values $J^*_{an}$: (a) 650°C; (b) 600°C.

Another possible cause for the acceleration of $J^*$ may be the small-scale creep condition [46], under which the creep-dominated region in the vicinity of the crack tip is confined and strained by the surrounding elastic-singularity-dominated region. However, at the present stage of research, no experimental evidence supporting the existence of the small-scale creep effect in creep–fatigue crack growth is available. To examine the possibility of this effect, therefore, finite element simulations were conducted on the crack growth in center-cracked plates under creep–fatigue conditions [47]. In the simulations, trapezoidal stress waves were employed, and the $J^*$ values were calculated by line integral techniques.

Figure 13(a) shows the variation of $J^*$ with time $t$ from the start of the maximum load hold. The paths used in the calculation of $J^*$ are shown in Fig. 13(b), the path numbers given in Fig. 13(b) corresponding to the numbers in Fig. 13(a). Considerations on stress and strain under small-scale creep conditions lead to the conclusion [46] that the $J^*$ value in the vicinity of the tip of a stationary crack under constant load is given by

$$J^* = J_{el}/(n + 1)t \qquad (12)$$

(see Appendix II) where $J_{el}$ denotes the elastic $J$-integral. Equation (12) is represented by the chain line in Fig. 13, where $J_{el}$ is replaced by the elastic $J$-integral range $\Delta J_{el}$. From the figure, the small-scale creep solution for a stationary crack expressed by Eqn. (12) gives some qualitative characteristics of the simulated behavior of $J^*$, the $J^*$ value in the vicinity of the crack tip (path number 1) being very high for small $t$ and decreasing in almost inverse proportion to $t$. Under the small-scale creep condition, the $J^*$ value is always higher than $J^*_{st}$, which represents the $J^*$ value under large-scale creep conditions. This figure suggests that

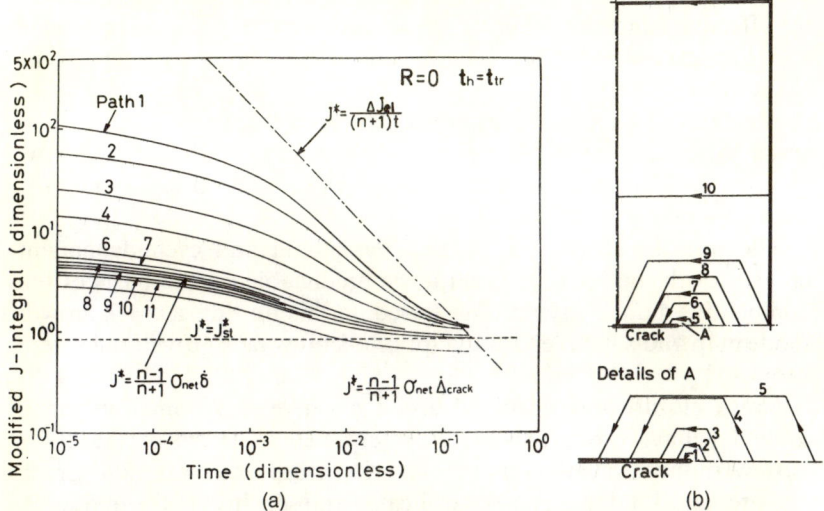

FIG. 13. Finite element simulation of the small-scale creep effect on $J^*$ for a center-cracked plate.

small-scale creep may cause accelerated creep deformation near the crack tip, which may result in accelerated crack growth.

The creep recovery which occurs during strain cycling may also lead to the acceleration of $J^*$ and was discussed elsewhere in some detail [48,49].

These $J^*$-accelerating factors should be kept in mind when estimating $da/dt$ from crack growth characteristics expressed in terms of $J^*$.

## CONCLUSIONS

The crack growth behavior of creep-ductile type 304 stainless steels under creep and creep–fatigue conditions has been discussed.

Creep crack growth rates $da/dt$ were best correlated with the modified $J$-integral $J^*$.

It was shown that the crack growth under creep–fatigue conditions was either purely time dependent or purely cycle dependent, depending on the test conditions employed.

The time-dependent crack growth rate was expressed solely as a function of the modified $J$-integral $J^*$, irrespective of the stress waveforms, their sequences and the loading conditions applied. Similarly, the cycle-dependent crack growth rate was uniquely characterized by the effective J-integral range $\Delta J_f$. Fracture surfaces were intergranular and transgranular for the time-dependent and the cycle-dependent crack growths, respectively.

A threshold for crack growth was observed in the cycle-dependent crack growth under creep–fatigue conditions. In the region of $\Delta J_f$ below this threshold, time-dependent crack growth appeared, behaving as if 'pseudo-thresholds' existed.

The transition between the time-dependent and cycle-dependent crack growth modes was found to be predictable on the basis of the competitive damage hypothesis. Based on this finding, a crack growth mode map and a simple formula for predicting crack growth rates were proposed.

Crack closure was observed under negative $R$. Anomalous crack growth behavior was reasonably interpreted in terms of the crack closure induced presumably by plasticity, creep and oxidation. Crack closure was found to be very complicated and sensitive to the frequency and loading history.

Some factors contributing to the acceleration of $J^*$ were discussed, which may lead to accelerated time-dependent crack growth.

# REFERENCES

[1]   M. J. Siverns and A. T. Price, *Nature,* **228**, 760 (1970).
[2]   S. Kubo, K. Ohji and K. Ogura, *Sci. Machine,* **28**, 1397 (1976); **29**, 27 (1977); **29**, 266 (1977).
[3]   K. Ohji, *Theor. Appl. Mech.,* **27**, 3 (1977).
[4]   H. P. Van Leeuwen, *Eng. Fract. Mech.,* **9**, 951 (1977).
[5]   E. G. Ellison and M. P. Harper, *J. Strain Anal.,* **13**, 35 (1978).
[6]   R. Pilkington, *Mater. Sci.,* **13**, 555 (1979).
[7]   L. S. Fu, *Eng. Fract. Mech.,* **13**, 307 (1980).
[8]   M. H. El Haddad, T. H. Topper and B. Mukherjee, *J. Test. Eval.,* **9**, 65 (1981).
[9]   K. Sadananda and P. Shahinian, *Eng. Fract. Mech.,* **15**, 327 (1981).
[10]  S. N. Malik, *Nucl. Eng. Design,* **72**, 359 (1982).
[11]  E. G. Ellison and G. G. Musicco, *Subcritical Crack Growth due to Fatigue, Stress Corrosion and Creep,* L. H. Larsson, Ed., Elsevier, Amsterdam, 403 (1984).
[12]  A. Saxena and V. Kumar, *J. Test. Eval.,* **12**, 189 (1984).
[13]  K. Sadananda and P. Shahinian, *Creep-Fatigue-Environment Interaction,* R. M. Pelloux and N. S. Stoloff, Eds., Metall. Soc. AIME, 86 (1980).
[14]  K. Ohji, *Proc. 28th Jpn. Natl. Symp. on Strength, Fracture and Fatigue,* 31 (1983).
[15]  G. J. Lloyd, *Fatigue at High Temperature,* R. P. Skelton, Ed., Applied Science, Barking, 187 (1983).
[16]  R. Ohtani, *Bull. Jpn. Inst. Met.,* **22**, 190 (1983).
[17]  E. G. Ellison, *Subcritical Crack Growth due to Fatigue, Stress Corrosion and Creep,* L. H. Larsson, Ed., Elsevier, Amsterdam, 531 (1984).
[18]  K. Sadananda, *Nucl. Eng. Design,* **83**, 303 (1984).
[19]  K. Ohji, *Role of Fracture Mechanics in Modern Technology: Proc. Int. Conf., Fukuoka, Japan,* North Holland, Amsterdam, 131 (1987).
[20]  K. Ohji, K. Ogura and S. Kubo, *Preprint of Jpn. Soc. Mech. Engrs.,* No. 740-11, 207 (1974).
[21]  K. Ohji, K. Ogura and S. Kubo, *Trans. Jpn. Soc. Mech. Eng.,* **42**, 350 (1976).
[22]  J. D. Landes and J. A. Begley, *ASTM STP 590,* 128 (1976).
[23]  K. M. Nikbin, G.A. Webster and C.E. Turner, *ASTM STP 601,* 47 (1976).
[24]  S. Kubo, Study on Mechanics of Creep Crack Initiation and Growth, *Doctoral Thesis,* Osaka University, (1975).
[25]  K. Ohji, K. Ogura and S. Kubo, *Trans. Jpn. Soc. Mech. Eng.,* **44**, 1831 (1978).
[26]  K. Ohji, K. Ogura and S. Kubo, *Trans. Jpn. Soc. Mech. Eng., Ser.A,* **47**, 400 (1981).
[27]  S. Taira, K. Tanaka and R. Koterazawa, *Proc. 2nd Japan Congr. on Testing of Materials,* Society of Materials Science, Kyoto, 55 (1959).
[28]  N. E. Dowling, *ASTM STP 601,* 19 (1976).
[29]  K. Ohji, K. Ogura and Y. Katada, *Preprint of 25th Annual Meeting of Jpn. Soc. Mater. Sci.,* Society of Materials Science, Kyoto, 19 (1976).
[30]  K. Ohji, K. Ogura, S. Kubo and Y. Katada, *Proc. 21st Jpn. Congr. on Materials Research,* Society of Materials Science, Kyoto, 99 (1978).

[31] K. Ohji, K. Ogura, Y. Katada, M. Takemoto and H. Nakajima, *J. Soc. Mater. Sci., Jpn.,* **27**, 1089 (1978).

[32] K. Ohji, K. Ogura, S. Kubo, Y. Katada and H. Saito, *J. Soc. Mater. Sci., Jpn.,* **27**, 1165 (1978).

[33] Y. Katada, Experimental Study on Creep Crack Growth Behaviour, *Doctoral Thesis,* Osaka University (1979).

[34] K. Ohji, K. Ogura, S. Kubo, Y. Katada, T. Katsuhara and N. Iwanaga, *Trans. Jpn. Soc. Mech. Eng., Ser.A,* **45**, 550 (1979).

[35] K. Ohji, K. Ogura, S. Kubo and Y. Katada, *Proc. Int. Conf. on Engineering Aspects of Creep, IMechE,* Institute of Mechanical Engineers, London, **2**, 9 (1980).

[36] K. Ohji, S. Kubo, N. Yamakawa and T. Kuri, *Trans. Jpn. Soc. Mech. Eng., Ser.A,* **50**, 1218 (1984).

[37] K. Ohji, S. Kubo, W. S. Joo and T. Kuri, *Trans. Jpn. Soc. Mech. Eng., Ser.A,* **51**, 2137 (1985).

[38] K. Ohji, S. Kubo, T. Kuri and W. S. Joo, *Proc. Int. Conf. on Creep,* Jpn. Soc. Mech. Eng., Tokyo, 297 (1986).

[39] K. Ohji, K. Ogura, S. Kubo, H. Saito and M. Fukumoto, *J. Mater. Sci., Jpn.,* **33**, 145 (1984).

[40] K. Ohji, S. Kubo, W. S. Joo, N. Yamakawa and T. Kuri, *J. Soc. Mater. Sci., Jpn.,* **34**, 692 (1985).

[41] K. Ohji, S. Kubo, T. Kuri and K. Nishizaki, *Trans. Jpn. Soc. Mech. Eng., Ser.A,* **51**, 2146 (1985).

[42] K. Ohji, S. Kubo and Y. Manabe, *J. Soc. Mater. Sci., Jpn.,* **35**, 924 (1986).

[43] K. Ohji, S. Kubo, T. Tokuhiro and S. Iwai, *Preprint of 63rd Natl. Meeting of Jpn. Soc. Mech. Eng.,* Jpn. Soc. Mech. Eng., Tokyo, 44 (1985).

[44] S. Taira, R. Ohtani and T. Komatsu, *Trans. ASME, Ser.H, J. Eng. Mater. Technol.,* **101**, 162 (1979).

[45] T. Kitamura and R. Ohtani, *Trans. Jpn. Soc. Mech. Eng., Ser.A,* **52**, 1816 (1986).

[46] K. Ohji, K. Ogura and S. Kubo, *J. Mater. Sci., Jpn.,* **29**, 465 (1980).

[47] K. Ohji, S. Kubo, T. Shibata and K. Ogura, *Trans. Jpn. Soc. Mech. Eng., Ser.A,* **51**, 90 (1985).

[48] S. Kubo, *ASTM STP 803,* I-594 (1983).

[49] S. Kubo, Brown University Report, MRL E-133 (1981).

[50] J. R. Rice, P. C. Paris and J. G. Merkle, *ASTM STP 536,* 231 (1973).

[51] J. W. Hutchinson, *J. Mech. Phys. Solids,* **16**, 13 (1968).

[52] J. R. Rice and G. F. Rosengren, *J. Mech. Phys. Solids,* **16**, 1 (1968).

[53] H. Riedel and J. R. Rice, *ASTM STP 700,* 112 (1980).

## APPENDIX I: EXPRESSION FOR THE EXPERIMENTAL EVALUATION OF $J^*$ [24, 25]

The modified $J$-integral $J^*$ was introduced by applying the analogue in the creep stress analysis to the $J$-integral. By referring to the energy rate interpretation of the $J$-integral [50], $J^*$ can be defined as

$$J^* = \frac{1}{B} \int \left(\frac{\partial \dot{\Delta}}{\partial a}\right)_P dP \qquad (13)$$

where $\dot{\Delta}$ is the load-point displacement rate, $B$ is the specimen thickness, $a$ is the crack length and $P$ is the applied load.

For a deep center-cracked specimen, the geometrical parameter which may affect $\dot{\Delta}$ is only the ligament length $b$ [50]. Then, by assuming the predominance of Norton-type creep, the following equation may be obtained by dimensional analysis:

$$\dot{\Delta} = C(P/Bb)^n b \qquad (14)$$

where $n$ denotes the creep exponent and $C$ is a constant.

Substitution of Eqn. (14) into Eqn. (13) yields

$$J^* = [(n-1)/(n+1)] \, \dot{\Delta}(P/Bb) = [(n-1)/(n+1)] \, \dot{\Delta}\sigma_{net} \qquad (15)$$

The factor $(n-1)/(n+1)$ in Eqn. (15) may be omitted, since it is nearly equal to unity for large $n$ values.

Equation (15) is valid for relatively deep cracks, and the range of applicability of this equation may be extended even to shallow cracks when $\dot{\Delta}$ is replaced by the crack-center opening rate [25].

## APPENDIX II: $J^*$ UNDER SMALL-SCALE CREEP CONDITIONS [46]

In situations where the creep-dominant region is limited within a near-tip region, the value of $J^*$ may be very high due to elastic straining by the surrounding elastic-singularity-dominated field. An approximate expression for $J^*$ was obtained as follows [46].

Let us assume that the total strain rate $\dot{\varepsilon}$ is given by the summation of the elastic component $\dot{\varepsilon}_{el}$ and the creep component $\dot{\varepsilon}_{cr}$, and that the latter is given by Norton's law:

$$\dot{\varepsilon}_{cr} = A\sigma^n \qquad (16)$$

Then, the strain at time $t$ is expressed as

$$\varepsilon = \varepsilon_{el} + \int_0^t A\,[\sigma(T)]^n \, dT = \sigma(t)/E + AF(t)[\sigma(t)]^n \qquad (17)$$

where $E$ denotes Young's modulus and $F(t)$ is defined as

$$F(t) = \int_0^t \left[\frac{\sigma(T)}{\sigma(t)}\right]^n dT \qquad (18)$$

We now assume that $F(t)$ evaluated in the vicinity of the crack tip can be applied to the whole body. This assumption may cause a small error in Eqn. (17). However, since the second term in Eqn. (17) is relatively small in the region far from the crack tip when compared with the first term, its effect may be negligibly small in evaluating the stress and strain fields in the vicinity of the creep crack tip. In this case, Eqn. (17) is formally taken to be equivalent to a non-linear stress–strain relationship. Hence, referring to the crack-tip solution for nonlinear elastic materials [51, 52], it is concluded that the path independence of the $J$-integral holds and that the Hutchinson–Rice–Rosengren (HRR) singular fields prevail in the near-tip region:

$$\sigma_{ij} = [J/IAF(t)r]^{1/(n+1)} \tilde{\sigma}_{ij}(\theta)$$
$$\varepsilon_{ij} = AF(t)[J/IAF(t)r]^{n/(n+1)} \tilde{\varepsilon}_{ij}(\theta) \tag{19}$$

where $r$ and $\theta$ are the polar coordinates with the origin at the crack tip, $\tilde{\sigma}_{ij}(\theta)$ and $\tilde{\varepsilon}_{ij}(\theta)$ are eigenfunctions of the singular fields and $I$ is a constant.

For the small-scale creep conditions under static load,

$$J = J_{el} = K^2/E \text{ (for plane stress)} = \text{constant} \tag{20}$$

Substituting Eqns. (19) and (20) into Eqn. (18), we obtain

$$F(t) = \int_0^t \left[ \frac{F(t)}{F(T)} \right]^{n/(n+1)} dT \tag{21}$$

A solution of Eqn. (21) is expressed as

$$F(t) = (n+1)t \tag{22}$$

Using Eqns. (19) and (22), the stress and strain rate near the crack tip are given as

$$\sigma_{ij} = [J_{el}/IA(n+1)tr]^{1/(n+1)} \tilde{\sigma}_{ij}(\theta)$$
$$\dot{\varepsilon}_{ij} = A[J_{el}/IA(n+1)tr]^{n/(n+1)} \tilde{\varepsilon}_{ij}(\theta) \tag{23}$$

On the other hand, $J^*$ can be defined when creep deformation is predominant in the near-tip region even under small-scale creep conditions. The near-tip fields are expressed in terms of $J^*$ as

$$\sigma_{ij} = [J^*/IAr]^{1/(n+1)} \tilde{\sigma}_{ij}(\theta)$$
$$\dot{\varepsilon}_{ij} = A[J^*/IAr]^{n/(n+1)} \tilde{\varepsilon}_{ij}(\theta) \tag{24}$$

By comparing Eqns. (23) and (24), an expression for $J^*$ under small-scale creep conditions is obtained:

$$J^* = J_{el}/(n + 1)t \tag{25}$$

The validity of Eqn. (25) has been confirmed by finite element calculations [46]. Essentially the same results were independently obtained by Riedel and Rice [53].

# Damage Evaluation and Life Prediction Under Creep-Fatigue Loading Condition for Austenitic Stainless Steels and Low Alloy Steel

KOICHI YAGI, KIYOSHI KUBO, OSAMU KANEMARU and CHIAKI TANAKA
*Creep Testing Division, National Research Institute for Metals,
2-3-12 Nakameguro, Meguro-ku, Tokyo, Japan*

## ABSTRACT

Life prediction for creep–fatigue loading conditions should be related to creep damage mechanisms. In order to examine the effect of the creep damage mode on rupture life under creep–fatigue loading, a combined creep–fatigue loading test was carried out on SUS 316, 304 and 1Cr–Mo–V steels. In this test method, the creep loading and fatigue loading were repeated alternately. The fracture criterion under the combined loading closely depended on the creep fracture mode of a static creep test. The fracture criteria obtained from the combined creep–fatigue loading tests agreed better with those obtained from fatigue tests with a tensile strain–hold wave form for SUS 316 steel than for 1Cr–Mo–V steel. A new life prediction method which uses this fracture criterion is proposed, the criteria being changed when the creep damage mode varies. In order to verify the adequacy of this method, fatigue tests with a tensile strain–hold wave form were carried out on SUS 316 steel, and the predicted life agreed well with the observed life.

## INTRODUCTION

The structural materials in high-temperature plants are frequently used under complicated conditions that include varying loads and temperatures. An understanding of the behavior of materials under such conditions is important for designing and maintaining components, and for establishing their safety and reliability [1]. The service conditions at the heated surfaces of heavy components can be simulated by loading conditions for which creep loading and fatigue

115

loading are alternately repeated at a constant temperature [2–5]. It is considered that under this loading condition the fatigue loading corresponds to the thermal loading which results from the temperature change accompanying the start-up and shut-down of a plant, and the creep loading corresponds to the residual stress resulting from this thermal loading.

Long-term creep–fatigue behavior must be considerably affected by material deterioration due to creep damage. The type of creep damage in a static creep test is a function of the temperature and stress [6–8]. Therefore, long-term creep–fatigue behavior must be understood in connection with the relevant creep damage mode. The creep fracture mode which is obtained from metallographic observations of ruptured specimens in a static creep test reflects the dominant damage mode during creep. If creep–fatigue behavior is examined under creep loading conditions which relate to a specific type of creep fracture mode, then the relationship between the creep damage mode and creep–fatigue interaction might be better understood.

Creep–fatigue behavior is generally examined by conducting a fatigue test with a tensile strain–hold wave form. However, this test method is not suitable for investigating the relationship between a fixed creep damage mode and creep–fatigue interaction because the stress is not constant during the tensile strain–hold. The authors have been systematically studying the effect of creep damage and fatigue damage on rupture life using a combined creep–fatigue loading test method [9–13], in which the creep loading and the fatigue loading are repeated alternately. It is considered that this test method is adequate for examining creep–fatigue interactions under given creep damage mode conditions. The loading schedule for this test method is also realistic as already described.

In this chapter a creep fracture mode map was first obtained experimentally for SUS 316 stainless steel. Then combined creep–fatigue loading tests were carried out under the creep loading condition corresponding to each creep fracture mode zone of the map, in order to examine the connection between the creep damage mode and creep–fatigue interaction. Furthermore, the relationship between the creep damage mode and creep–fatigue interaction was also examined for SUS 304 stainless steel and 1Cr-Mo-V steel. Fracture criteria under the creep–fatigue loading conditions obtained from the combined creep–fatigue loading tests were compared with those obtained from fatigue tests with a tensile strain–hold wave form for SUS 316 steel and 1Cr-

Mo–V steel, and the influence of the test method on creep–fatigue interaction was examined. A new life prediction method which takes account of the change in the creep damage mechanism is proposed from the damage evaluation of the combined creep–fatigue loading tests. This life prediction method is verified by fatigue tests with a tensile strain–hold wave form for SUS 316 steel.

## EXPERIMENTAL PROCEDURES

### Materials and Specimens

The materials tested were SUS 316 steel plate of thickness 24 mm, SUS 304 steel plate of thickness 21 mm, and 1Cr–Mo–V steel forgings for turbine rotors of outside diameter 1160 mm and 1190 mm, which are designated as 1Cr–Mo–V steel A and 1Cr–Mo–V steel B, respectively. The chemical composition of each material is shown in Table 1. The solution heat treatment condition of SUS 316 and 304 steels was 1100 °C for 20 min with water quenching and 1100 °C for 30 min with water quenching, respectively, the austenitic ASTM grain size of these steels being approximately 6. The heat treatment condition of 1Cr–Mo–V steels A and B is shown in Table 2, the prior austenitic ASTM grain size of each steel being 5·8 and 7·8, respectively.

The dimension of the specimens used for the combined creep–fatigue loading tests and fatigue tests was 6 mm in diameter over a parallel portion 13·2 mm in length, with edges at both ends of the parallel portion for fixing an extensometer. The portion between the edges for these tests was polished in the longitudinal direction after machining. The dimensions of the specimens used for the static creep tests were 10 mm in diameter for the parallel portion and 50 mm in gauge length. The axis of the specimens in SUS 316 and 304 steels was parallel with the rolling direction of the plates. The specimens in 1Cr–Mo–V steel A were machined from the radial direction of the rotor, and the specimens in 1Cr–Mo–V steel B were machined from the axial direction of the rotor.

### Testing Procedures and Testing Conditions

The static creep test and the strain-controlled fatigue test were carried out to obtain basic data which could be compared with the rupture life from combined creep–fatigue loading tests, and to produce the experimental creep fracture mode map. The static creep tests were

TABLE 1
Chemical composition of the materials (wt.%)

| Steel | C | Si | Mn | P | S | Ni | Cr | Mo | Cu | V | Sb | Ti | Al | B | N | Nb+Ta |
|-------|---|----|----|---|---|----|----|----|----|---|----|----|----|---|---|-------|
| SUS 316 | 0·05 | 0·70 | 1·10 | 0·034 | 0·003 | 12·60 | 17·05 | 2·24 | 0·31 | — | — | 0·03 | <0·003 | 0·003 | 0·017 | 0·001 |
| SUS 304 | 0·06 | 0·67 | 1·02 | 0·025 | 0·007 | 9·14 | 18·48 | 0·18 | 0·07 | — | — | 0·021 | <0·003 | — | 0·02 | — |
| 1Cr-Mo-V, A | 0·29 | 0·20 | 0·75 | 0·010 | 0·009 | 0·34 | 1·00 | 1·25 | 0·14 | 0·26 | — | — | <0·002 | — | 0·007 5 | — |
| 1Cr-Mo-V, B | 0·29 | 0·30 | 0·74 | 0·006 | 0·003 | 0·39 | 1·12 | 1·16 | 0·03 | 0·25 | 0·001 8 | — | <0·003 | — | 0·004 2 | — |

## TABLE 2
### Heat treatment condition of 1Cr–Mo–V steels

| 1Cr–Mo–V steel A | 1Cr–Mo–V steel B |
|---|---|
| 1010 °C × 25 h → 870 °C × 5·3 h → Air cooling | 1010 °C × 24 h → Air cooling |
| 720 °C × 38·5 h → Furnace cooling | 700 °C × 38 h → Air cooling |
| 960 °C × 17·5 h → Mist cooling | 650 °C × 5 h → 950 °C × 19 h → Air cooling |
| 665 °C × 65 h → Furnace cooling | 680 °C × 38 h → Furnace cooling |

carried out using a conventional lever-loaded type creep testing machine with a loading capacity of 30 kN. The strain-controlled fatigue tests were carried out on servo-hydraulic testing machines, with a strain rate of a triangular wave form to produce 6% min$^{-1}$.

The stress-controlled creep loading and strain-controlled fatigue loading were alternately repeated in the combined creep–fatigue loading test. The loading schedule for the stress and strain, and the hysteresis loops in this test are shown schematically in Fig. 1, the combined loading being continuously repeated until rupture of the specimen. The combined creep–fatigue loading tests were carried out on

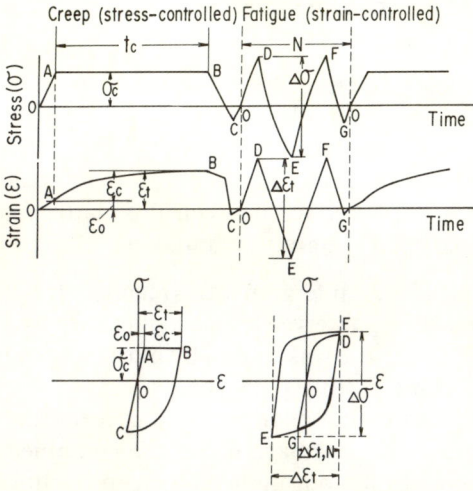

FIG. 1. Loading schedule for stress and strain, and hysteresis loops for the combined creep–fatigue loading test.

servo-hydraulic testing machines with two function generators for creep loading and for fatigue loading [14]. This combined creep–fatigue loading test method has the four parameters of creep stress $\sigma_c$ and creep time $t_c$ in creep loading, and total strain range $\Delta\varepsilon_t$ and number of cycles $N$ in fatigue loading. The value of $\sigma_c$ was chosen from the stress condition under which the time to rupture in a static creep test was about 1000 h at each test temperature, and under which the typical creep fracture mode was found. In each test, $t_c$ was 10 h with a few exceptions, and $\Delta\varepsilon_t$ was 1%. The strain rate under fatigue loading with a triangular wave form was 6% $\text{min}^{-1}$.

The fatigue tests with a tensile strain–hold wave form were carried out on SUS 316 steel and 1Cr–Mo–V steel B, with total strain ranges from 0·5% to 1·5%, and hold times of 0·1 and 1·0 h. The tension-going and compression-going strain rates were 6% $\text{min}^{-1}$.

## RESULTS AND DISCUSSIONS

### Creep Fracture Mode of SUS 316 Steel

In order to determine the creep loading conditions for the combined creep–fatigue loading test, the microstructure and fracture surface of the specimens ruptured in the static creep tests for SUS 316 steel were observed. Three kinds of creep fracture mode were recognized for this steel as shown in Fig. 2:

   (i) intergranular fracture due to the formation and growth of wedge-type intergranular cracks;
  (ii) transgranular fracture;
 (iii) intergranular fracture due to the formation and growth of cavities at the interface between the matrix and carbides or sigma phase on the grain boundaries.

As shown in Fig. 2, intergranular fracture due to wedge-type intergranular cracks was observed for the higher stresses at temperatures below 550 °C. This fracture was caused by hardening of the matrix due to strain aging during creep [11, 15, 16].

In order to clarify the relationship between the creep–fatigue interaction and the creep damage mode, the combined creep–fatigue loading tests were carried out under the creep loading conditions at which the typical creep fracture mode was observed. The arrows in Fig. 2 represent the creep loading conditions (temperature and stress) of

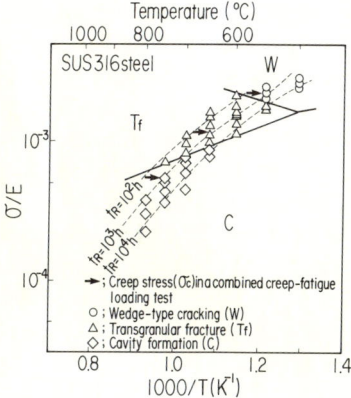

FIG. 2. Creep fracture mode map for SUS 316 steel.

the combined creep–fatigue loading tests. The creep stress $\sigma_c$ at 550 °C is within the region of intergranular fracture due to wedge-type intergranular cracks that were caused by strain aging, while $\sigma_c$ at 650 °C is within the region of transgranular fracture, and $\sigma_c$ at 750 °C is within the region of intergranular fracture due to the formation of cavities.

## Results of Combined Creep–Fatigue Loading Tests for SUS 316 Steel

Figure 3 and Fig. 4 respectively show the comparison of the total creep time to rupture in combined creep–fatigue loading tests, $\Sigma t_c$, with the time to rupture in static creep tests, and the comparison of the total number of cycles to rupture in fatigue loading from the combined

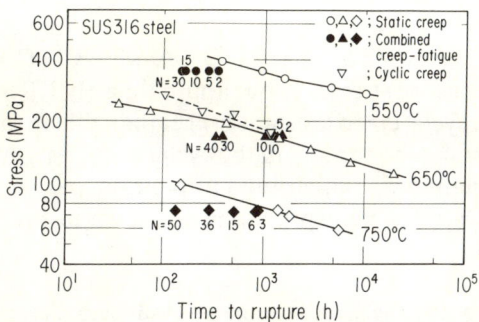

FIG. 3. Comparison of the time to rupture under combined creep–fatigue loadings with that under static creep loadings for SUS 316 steel.

creep–fatigue loading tests, $\Sigma N$, with the number of cycles to rupture in the strain-controlled fatigue tests. The time to rupture in the combined creep–fatigue loading tests was shorter than that in the static creep tests, except for the condition of $N \leqslant 10$ cycles at 650 °C, and decreased with increasing $N$ as shown in Fig. 3. The number of cycles to rupture in fatigue loading from the combined creep–fatigue loading tests was smaller than that in the strain-controlled fatigue tests, except for the condition of $N \geqslant 10$ cycles at 650 °C, and decreased with decreasing $N$ as shown in Fig. 4.

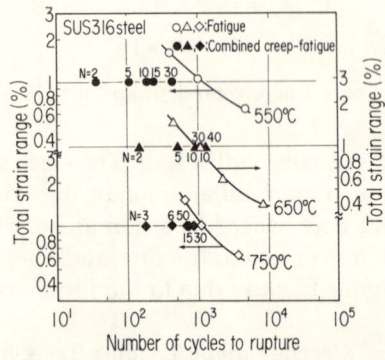

FIG. 4. Comparison of the number of cycles to rupture under combined creep–fatigue loadings with that under strain-controlled fatigue loadings for SUS 316 steel.

Figure 3 also shows the result of the cyclic creep tests at 650 °C. Only the creep loading portion of the combined creep–fatigue loading tests was repeated in the cyclic creep tests ($O \rightarrow A \rightarrow B \rightarrow C \rightarrow O \rightarrow A \rightarrow B \rightarrow \ldots$ in Fig. 1). In the case of stresses above approximately 170 MPa, the time to rupture in the cyclic creep tests was longer than that in the static creep tests. This longer time to rupture in the cyclic creep tests at 650 °C arose from the test method, in which the creep strain was suppressed and reduced to zero after the creep loading as shown in Fig. 1.

## Relationship Between the Creep Damage Mode and Creep–Fatigue Interaction for SUS 316 Steel

The creep damage $\phi_c$ and fatigue damage $\phi_f$ which were accumulated until rupture of the specimens in the combined creep–fatigue loading tests were calculated using the linear life fraction damage rule [17] as follows:

$$\phi_c = N_c \frac{t_c}{t_R} \tag{1}$$

$$\phi_f = N_c \left( \frac{N-1}{N_f} + \frac{1}{N_{ff}} \right) \tag{2}$$

where $N_c$ is the number of combined cycles to rupture in a combined creep–fatigue loading test, $t_R$ is the time to rupture in a static creep test under the loading condition of $\sigma_c$, $N_f$ is the number of cycles to rupture in a strain-controlled fatigue test under the total strain range condition of $\Delta\varepsilon_t$, and $N_{ff}$ is the number of cycles to rupture in the fatigue test at $\Delta\varepsilon_{t,N}$, which is shown in Fig. 1. The value of $\Delta\varepsilon_{t,N}$ corresponds to the strain range at the $N$th cycle in fatigue loading during the combined creep–fatigue loading test.

Figure 5 shows the relation between $\phi_c$ and $\phi_f$ calculated from Eqns. (1) and (2). Three types of $\phi_c$ versus $\phi_f$ relation were found, and the connection between the $\phi_c$ versus $\phi_f$ relation and the creep loading conditions was examined. This connection may be summarized as follows.

(i) In the case of the creep loading condition within the region of the wedge-type intergranular fracture mode caused by strain aging, the $\phi_c$ versus $\phi_f$ relation was

$$\phi_c + 2\cdot66(\phi_c\phi_f)^{1/2} + \phi_f = 1 \tag{3}$$

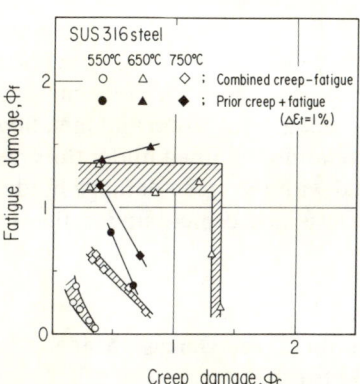

FIG. 5. Relations of creep damage $\phi_c$ to fatigue damage $\phi_f$, calculated according to the rupture life under combined creep–fatigue loadings for SUS 316 steel using the linear life fraction damage rule.

(ii) In the case of the creep loading condition within the transgranular fracture mode region, the $\phi_c$ versus $\phi_f$ relation was

$$\phi_c = 1 \text{ to } 2 \text{ [9]} \quad \text{or} \quad \phi_f \approx 1 \tag{4}$$

(iii) In the case of the creep loading condition within the cavity-type intergranular fracture mode region, the $\phi_c$ versus $\phi_f$ relation was

$$\phi_c + \phi_f \approx 1 \tag{5}$$

The effect of prior creep loading on the fatigue life was examined, the creep damage when subjected to prior creep loadings and the fatigue damage when subjected to the strain-controlled fatigue tests for the crept materials being calculated as follows:

$$\phi_c = \frac{t_A}{t_R} \tag{6}$$

$$\phi_f = \frac{N_{Af}}{N_f} \tag{7}$$

where $t_A$ is the prior creep time in static creep tests under the creep loading condition of $\sigma_c$, and $N_{Af}$ is the number of cycles to rupture in a fatigue test at $\Delta\varepsilon_t$ of 1% for the prior crept materials. The results are also shown by the solid symbols in Fig. 5. The $\phi_c$ versus $\phi_f$ relations obtained from the fatigue tests for prior crept materials show the same trends as those obtained from the combined creep–fatigue loading tests. The fracture modes observed in the combined creep–fatigue loading tests coincided with those observed in the fatigue tests on the prior crept materials [11]. It was also found that the intergranular damage mode under combined creep–fatigue loadings coincided with the creep damage mode observed for the prior crept materials. It is concluded from the results of a damage evaluation and metallographic observation that the $\phi_c$ versus $\phi_f$ relation obtained under the creep–fatigue loadings was closely connected with the creep damage mode and fracture mode of the static creep tests performed under the same creep loading conditions.

## Relationship Between the Creep Damage Mode and $\phi_c$ Versus $\phi_f$ Relation for SUS 304 Steel

Figure 6 shows the creep fracture mode in static creep tests for SUS 304 steel. In order to clarify the relationship between the creep damage mode and $\phi_c$ versus $\phi_f$ relation, the combined creep–fatigue loading tests were carried out under the creep loading conditions which are represented by the arrows in Fig. 6. The creep stress $\sigma_c$ at 600 °C and that for higher stresses at 700 °C are within the region of transgranular fracture, and $\sigma_c$ for the lower stresses at 700 °C is within the region of intergranular fracture due to the formation of cavities and wedge-type cracking.

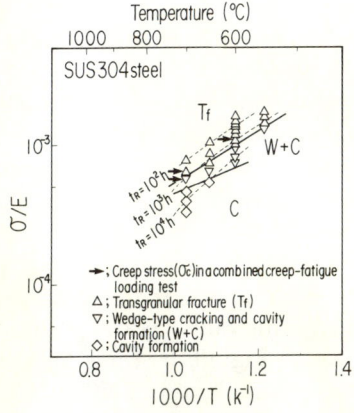

| T(°C) | $\sigma_c$(MPa) | $t_c$(h) | $\Delta\varepsilon_t$(%) | N(cycles) |
|---|---|---|---|---|
| ○ ; 600 | 177 | 10 | 1 | 2 – 20 |
| ◇ ; 700 | 98 | 5 | 1 | 5 , 10 |
| △ ; 700 | 86 | 10 | 1 | 2 – 20 |

FIG. 6. Creep fracture mode map for SUS 304 steel.

FIG. 7. Relation of creep damage $\phi_c$ to fatigue damage $\phi_f$, obtained from combined creep–fatigue loading tests for SUS 304 steel.

Figure 7 shows the $\phi_c$ versus $\phi_f$ relation calculated from Eqns. (1) and (2). Two types of $\phi_c$ versus $\phi_f$ relation were found for SUS 304 steel, and the connection between the $\phi_c$ versus $\phi_f$ relation and the creep damage mode was examined. In the case of creep loading conditions within the transgranular fracture mode region, the $\phi_c$ versus $\phi_f$ relation was expressed by Eqn. (4), and in the case of creep loading conditions within the intergranular fracture mode region due to the formation of cavities and wedge-type cracking, the $\phi_c$ versus $\phi_f$ relation was expressed by Eqn. (5). The relationship between the creep damage mode and the $\phi_c$ versus $\phi_f$ relation for SUS 304 steel was the same as that for SUS 316 steel, and is summarized in Fig. 8.

### Relationship Between the Creep Damage Mode and $\phi_c$ Versus $\phi_f$ Relation for 1Cr–Mo–V Steel

SUS 316 and 304 steels are cyclic-strain hardening materials. In order to examine the relationship between the creep damage mode and $\phi_c$

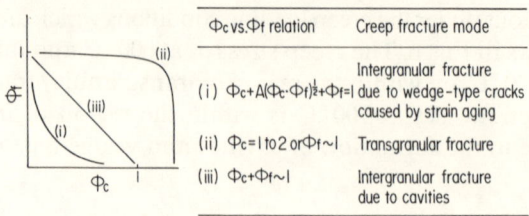

| $\Phi_c$ vs. $\Phi_f$ relation | Creep fracture mode |
|---|---|
| (i) $\Phi_c + A(\Phi_c \cdot \Phi_f)^{\frac{1}{2}} + \Phi_f = 1$ | Intergranular fracture due to wedge-type cracks caused by strain aging |
| (ii) $\Phi_c = 1$ to 2 or $\Phi_f \sim 1$ | Transgranular fracture |
| (iii) $\Phi_c + \Phi_f \sim 1$ | Intergranular fracture due to cavities |

FIG. 8. Relationship between $\phi_c$ versus $\phi_f$ relations and creep fracture modes.

versus $\phi_f$ relation for a cyclic-strain softening material, combined creep-fatigue loading tests were carried out for 1Cr–Mo–V steel A. The creep fracture mode in static creep tests for 1Cr–Mo–V steel A was transgranular within the limits of this experiment as shown in Fig. 9 [13]. The combined creep–fatigue loading tests were carried out under the creep loading conditions shown by two arrows in Fig. 9. The creep damage $\phi_c$ and the fatigue damage $\phi_f$ accumulated until rupture in the combined creep–fatigue loading tests were evaluated using Eqns. (1) and (2), the $\phi_c$ versus $\phi_f$ relation obtained being shown in Fig. 10. It seems that the rupture life could be determined by the accumulation of either creep damage or fatigue damage. Figure 10 also shows the results of creep-fatigue interspersion tests conducted by the Metal Properties Council [4], the $\phi_c$ versus $\phi_f$ relation obtained from the present test results agreeing well with those obtained by the Metal Properties Council.

Figure 11 shows a comparison of the $\phi_c$ versus $\phi_f$ relation for 1Cr–Mo–V steel with those for SUS 316 and 304 steels, when the combined creep–fatigue loading tests were carried out under creep

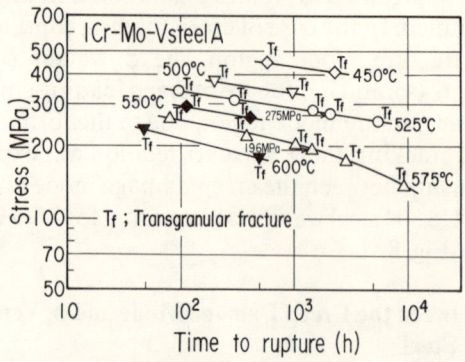

FIG. 9. Time to rupture and fracture mode in static creep rupture tests for 1Cr–Mo–V steel A.

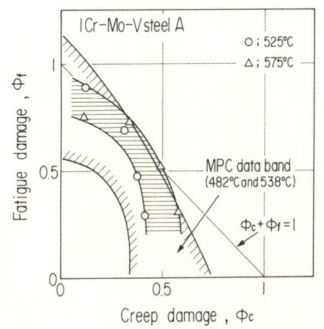

FIG. 10. Relation of creep damage $\phi_c$ to fatigue damage $\phi_f$, obtained from combined creep-fatigue loading tests for 1Cr-Mo-V steel A.

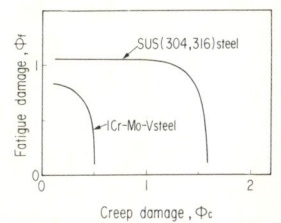

FIG. 11. Comparison of the $\phi_c$ versus $\phi_f$ relation obtained from combined creep-fatigue loading tests on 1Cr-Mo-V steel with that obtained on SUS 304 and 316 steels, when the creep loading condition corresponds to the transgranular fracture region on a creep fracture mode map.

loading conditions corresponding to the region of the transgranular fracture mode in the static creep tests. The value of $\phi_c$ accumulated until rupture was approximately 0·5 for 1Cr-Mo-V steel, and 1 to 2 for SUS 316 and 304 steels, the critical value for creep damage permitted to rupture being smaller for 1Cr-Mo-V steel. This was caused by the decrease of creep resistance due to fatigue loading [13].

### Effect of Testing Procedures on the $\phi_c$ Versus $\phi_f$ Relation

In order to examine the effect of testing procedures on the $\phi_c$ versus $\phi_f$ relation, fatigue tests with a tensile strain–hold wave form were carried out on SUS 316 steel and 1Cr-Mo-V steel. Figure 12 and Fig. 13 show the test results for SUS 316 steel and 1Cr-Mo-V steel, respectively. The number of cycles to rupture decreased with increasing hold time $t_H$ and the degree of decrease was dependent on the material, temperature and total strain range.

The creep damage $\phi_c$ in the fatigue tests with a tensile strain–hold wave form was evaluated using the stress relaxation curve at half of the number of cycles to rupture as follows:

$$\phi_c = N_{fh} \int_0^{t_H} \frac{dt}{t_R} \qquad (8)$$

and the fatigue damage $\phi_f$ was evaluated from the equation

$$\phi_f = \frac{N_{fh}}{N_h} \qquad (9)$$

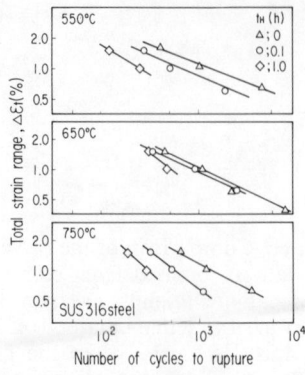

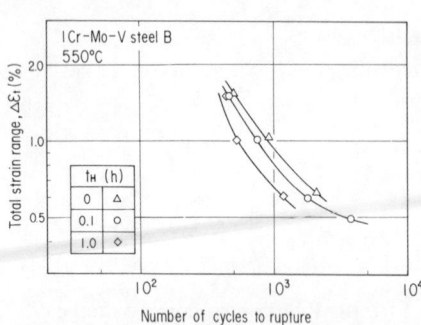

FIG. 12. Results of fatigue tests with a tensile strain–hold wave form for SUS 316 steel.

FIG. 13. Results of fatigue tests with a tensile strain–hold wave form for 1Cr-Mo-V steel B.

where $t_H$ is the hold time and $N_{fh}$ is the number of cycles to rupture in fatigue tests with a tensile strain–hold wave form. Figures 14(a) and 14(b) show the $\phi_c$ versus $\phi_f$ relation in fatigue tests with a tensile strain–hold wave form calculated from Eqns. (8) and (9) for SUS 316 steel and 1Cr-Mo-V steel, respectively. Figure 14 also shows the $\phi_c$ versus $\phi_f$ relation obtained from the combined creep–fatigue loading tests, which has already been represented in Figs. 5 and 10. The stress value during strain–hold in the fatigue tests with a tensile strain–hold wave form at 550 °C for SUS 316 steel was within the wedge-type cracking region in the creep fracture mode map of Fig. 2, and the stress value at 650 °C was within the transgranular fracture region. The stress value during strain–

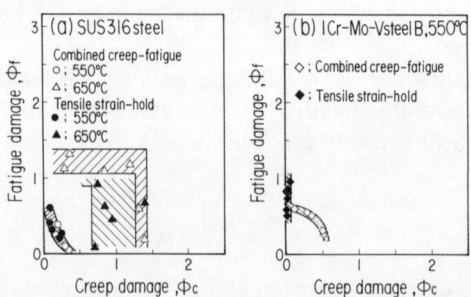

FIG. 14. Comparison of the $\phi_c$ versus $\phi_f$ relation obtained from combined creep–fatigue loading tests with that obtained from fatigue tests with a tensile strain–hold wave form for (a) SUS 316 steel and (b) 1Cr-Mo-V steel B.

hold for 1Cr-Mo-V steel was within the transgranular fracture region. For SUS 316 steel, the $\phi_c$ versus $\phi_f$ relation obtained from the fatigue tests with a tensile strain-hold wave form almost agreed with that obtained from the combined creep-fatigue loading tests under the creep loading conditions with the same creep damage mode. However, for 1Cr-Mo-V steel, the $\phi_c$ versus $\phi_f$ relation obtained from the fatigue tests with a tensile strain-hold wave form showed a different tendency from that obtained from the combined creep-fatigue loading tests, and $\phi_c$ in the fatigue tests with a tensile strain-hold wave form was smaller than that in the combined creep-fatigue loading tests. It is considered that the decrease in number of cycles to rupture with hold time for 1Cr-Mo-V steel was not caused by the accumulation of creep damage. $2\frac{1}{4}$Cr-1Mo steel also showed the same relation as 1Cr-Mo-V steel [18]. 1Cr-Mo-V steel and $2\frac{1}{4}$Cr-1Mo steel at high temperatures are more sensitive to oxidation than SUS 316 steel. Therefore, it seems that the decrease of rupture life in the fatigue test with a tensile strain-hold wave form for 1Cr-Mo-V steel was due to oxidation.

## LIFE PREDICTION TAKING ACCOUNT OF THE CREEP DAMAGE MODE

### Life Prediction Method

It was apparent from the combined creep-fatigue loading test results that the fracture criteria should be changed with the variation of creep damage mode. Figure 15 shows a procedure for life prediction taking account of variations in the creep damage mode. In the present work, the life under fatigue loading with a tensile strain-hold wave form for SUS 316 steel was predicted. For this fatigue test, the stress changed during the tensile strain-hold, so the material was subjected to various types of creep damage. The creep damage mode which was mainly formed at the stress level during relaxation could be identified from the creep fracture mode map in Fig. 2. The fracture criterion corresponding to each creep damage mode was determined from an evaluation of the combined creep-fatigue loading test result, each creep damage in one cycle of the tensile strain-hold wave form being evaluated as follows:

$$\Delta\phi_{cW} = \int_0^{t_H} \frac{dt}{t_{RW}} \tag{10}$$

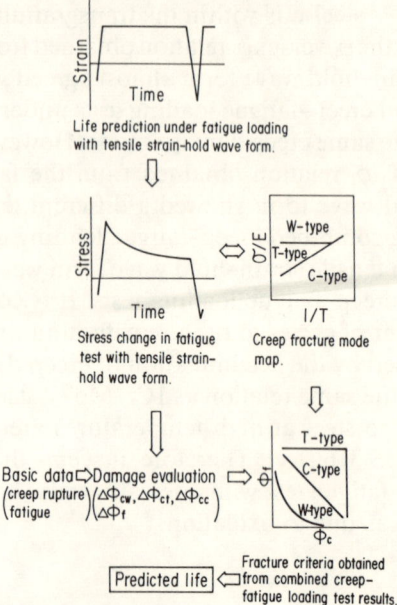

FIG. 15. Procedure for predicting the life under creep–fatigue conditions.

$$\Delta\phi_{cT} = \int_0^{t_H} \frac{dt}{t_{RT}} \tag{11}$$

$$\Delta\phi_{cC} = \int_0^{t_H} \frac{dt}{t_{RC}} \tag{12}$$

where $\Delta\phi_{cw}$ is the creep damage due to the wedge-type intergranular cracks caused by strain aging, $t_{RW}$ is the time to rupture in a static creep test under the creep damage mechanism, $\Delta\phi_{cT}$ is the creep damage for which the formation of intergranular cracks is very slow [19] and the fracture in a static creep test is transgranular, $t_{RT}$ is the time to rupture in a static creep test under the creep damage mechanism, $\Delta\phi_{cC}$ is the creep damage due to cavity-type intergranular cracks, and $t_{RC}$ is the time to rupture in a static creep test under the creep damage mechanism. In the present work, the creep damage was evaluated using the stress relaxation curve at half of the number of cycles to rupture in the fatigue tests with a tensile strain–hold wave form. The fatigue damage $\Delta\phi_f$ was evaluated as follows:

$$\Delta\phi_f = \frac{1}{N_F} \tag{13}$$

where $N_F$ is the number of cycles to rupture in the fatigue test with a triangular wave form in the same total strain range.

The fracture criteria under creep–fatigue loading for SUS 316 steel are defined on the basis of an evaluation of the experimental results as follows.

(i) In the case of a creep damage mode of the W type evaluated from Eqn. (10),

$$\{\Delta\phi_{cW} + 2\cdot66(\Delta\phi_{cW}\Delta\phi_f)^{1/2} + \Delta\phi_f\}\,N_{fpred} = 1 \tag{14}$$

(ii) In the case of a creep damage mode of the T type evaluated from Eqn. (11),

$$\Delta\phi_{cT}N_{fpred} = 1 \qquad \text{(creep dominant)} \tag{15}$$

or

$$\Delta\phi_f N_{fpred} = 1 \qquad \text{(fatigue dominant)} \tag{16}$$

(iii) In the case of a creep damage mode of the C type evaluated from Eqn. (12),

$$(\Delta\phi_{cC} + \Delta\phi_f)\,N_{fpred} = 1 \tag{17}$$

where $N_{fpred}$ is the life predicted by each fracture criterion. The fracture criterion which has a minimum of $N_{fpred}$ dominates the rupture life, and the minimum is the predicted life.

## Prediction of Rupture Life Under Fatigue Loading with a Tensile Strain–Hold Period

In this life prediction method, the time to rupture caused by each creep damage mechanism under static creep loading is necessary. However, it is difficult to obtain theoretically the accurate time to rupture for each creep damage mechanism. The functions of time to rupture, stress and temperature for a fixed creep damage mode can be experimentally found from the creep fracture mode map. The time-to-rupture–stress–temperature relationships corresponding to each creep damage mode were obtained as follows:

$$T(19 \cdot 651 + \log t_{RW}) = 31\ 401 - 5\ 024 \cdot 5 \log \sigma \qquad (18)$$

$$T(24 \cdot 085 + \log t_{RT}) = 43\ 321 - 8\ 237 \cdot 2 \log \sigma \qquad (19)$$

$$T(17 \cdot 353 + \log t_{RC}) = 31\ 639 - 5\ 732 \cdot 7 \log \sigma \qquad (20)$$

where $T$ is the temperature in K and $\sigma$ is the stress in MPa. The solid, broken and chain lines in Fig. 16 show the stress versus time to rupture curves expressed by Eqns. (18), (19) and (20), respectively. These curves agree well with the experimental data.

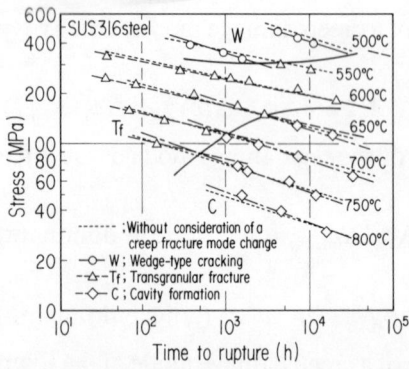

FIG. 16. Relationship between the stress and time to rupture for static creep tests in each creep fracture mode region for SUS 316 steel.

The creep damage corresponding to each creep damage mode during a tensile strain–hold was calculated by substituting the time to rupture obtained from Eqns. (18)–(20) into Eqns. (10)–(12), respectively. $N_{fpred}$ of each fracture criterion was calculated using Eqns. (14)–(17), the minimum $N_{fpred}$ value being the predicted life. Figure 17 shows a comparison of the predicted rupture life with the observed values, the predicted life agreeing well with the observed life within a factor of 2. The proposed new life prediction method was compared with a conventional method, in which the fracture criterion was not dependent on the creep damage mode. The life predicted by the proposed method for which the fracture criteria were changed by variation in the creep damage mode had a reduced scattering and less deviation [20].

The fracture criterion used to predict the life corresponded well with the observed fracture mode of the specimens ruptured in the fatigue test with a tensile strain–hold wave form [20]. It is suggested that this life

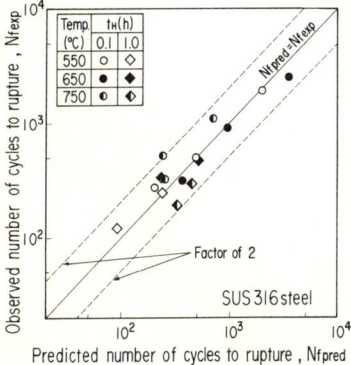

FIG. 17. Comparison of the predicted rupture life with the observed rupture life.

prediction method can also predict the fracture mode. It is clear from this investigation that the proposed life prediction method is more accurate and reasonable than the conventional method.

## CONCLUSIONS

In order to clarify the effect of the creep damage mode on the creep–fatigue interaction behavior, combined creep–fatigue loading tests were carried out for SUS 316 steel, SUS 304 steel and 1Cr–Mo–V steel. The creep damage $\phi_c$ and fatigue damage $\phi_f$ accumulated until rupture were evaluated by the linear life fraction damage rule, and the connection between the creep damage mode and the $\phi_c$ versus $\phi_f$ relation was examined. A new life prediction method which considered the relevant fracture criteria was proposed. The results can be summarized as follows.

(1) The $\phi_c$ versus $\phi_f$ relations obtained from combined creep–fatigue loading tests were closely connected with the creep fracture mode observed in the ruptured specimen from a static creep test under the same creep loading conditions.

(2) The fracture criteria under combined creep–fatigue loadings for SUS 316 steel were the same relations as those for SUS 304 steel. However, the critical value of creep damage permitted to rupture for 1Cr–Mo–V steel was smaller than that for SUS 316 and 304 steels.

(3) The $\phi_c$ versus $\phi_f$ relations obtained from the fatigue tests with a

tensile strain–hold wave form coincided well with those obtained from the combined creep–fatigue loading tests under the same creep damage mode conditions for SUS 316 steel. However, for 1Cr–Mo–V steel, the $\phi_c$ versus $\phi_f$ relation obtained from the fatigue tests with a tensile strain–hold wave form showed a different tendency from that obtained from the combined creep–fatigue loading tests. The decrease of rupture life in the fatigue tests with a tensile strain–hold wave form for 1Cr–Mo–V steel seemed to be dependent on oxidation.

(4) In the new life prediction method, the fracture criteria under creep–fatigue loading were changed according to the variation in creep damage mode, the rupture life predicted by the new method agreeing well with observed data. This method was more accurate than the conventional method without any consideration of the variation of creep damage modes.

(5) The new method also predicted the fracture mode under creep–fatigue loading.

# REFERENCES

[1]  *ASME Boiler and Pressure Vessel Code*, Section III, Division 1, Case N47-17 (1979).
[2]  D. P. Timo, *Proc. Int. Conf. on Thermal Stress and Thermal Fatigue*, CEGB, Berkeley, 439 (1969).
[3]  K. H. Kloos, J. Granacher and P. Rieth, *Arch. Eisenhüttenw.*, **52**, 237 (1981).
[4]  R. M. Curran and B. M. Wundt, *Proc. 1976 ASME–MPC Symp. on Creep-Fatigue Interaction*, ASME, New York, 203 (1976).
[5]  H. Breitling, E. D. Grosser and H. Lorenz, *Arch. Eisenhüttenw.*, **48**, 403 (1977).
[6]  R. J. Fields, T. Weerasooriya and M. F. Ashby, *Metall. Trans. A*, **11A**, 333 (1980).
[7]  N. Shinya, H. Tanaka and S. Yokoi, *Proc. 2nd Int. Conf. on Creep and Fracture of Engineering Materials and Structures*, Pineridge Press, Swansea, 739 (1984).
[8]  D. A. Miller and T. G. Langdon, *Metall. Trans. A*, **10A**, 1635 (1979).
[9]  K. Yagi, C. Tanaka and K. Kubo, *J. Soc. Mater. Sci. Jpn.*, **33**, 1078 (1984).
[10]  K. Yagi, C. Tanaka and K. Kubo, *J. Soc. Mater. Sci. Jpn.*, **33**, 1533 (1984).
[11]  K. Yagi, C. Tanaka, K. Kubo and O. Kanemaru, *Trans. Iron Steel Inst. Jpn.*, **25**, 1179 (1985).
[12]  K. Yagi, K. Kubo, O. Kanemaru and C. Tanaka, *J. Soc. Mater. Sci. Jpn.*, **34**, 1333 (1985).
[13]  K. Yagi, K. Kubo, O. Kanemaru and C. Tanaka, *J. Soc. Mater. Sci. Jpn.*, **35**, 434 (1986).
[14]  K. Yagi, C. Tanaka and K. Kubo, *J. Mater. Test. Res. Assoc. Jpn.*, **30**, 53 (1985).

[15]   J. T. Barnby, *J. Iron Steel Inst.*, **204**, 23 (1966).
[16]   D. G. Morris and D. R. Harries, *Met. Sci.*, **12**, 525 (1978).
[17]   S. Taira, *Creep in Structures*, N. J. Hoff, Ed., Springer, Berlin, 96 (1962).
[18]   K. Kubo, K. Yagi, T. Kaneko and O. Umezawa, unpublished work.
[19]   K. Yagi, K. Kubo, O. Kanemaru and C. Tanaka, *J. Soc. Mater. Sci. Jpn.*, **34**, 1082 (1985).
[20]   K. Yagi, O. Kanemaru, K. Kubo, and C. Tanaka, *J. Soc. Mater. Sci. Jpn.*, **35**, 176 (1986).

# Effect of Environmental Air on the Creep–Fatigue Fracture Behavior of High Temperature Structural Materials

KATSUYUKI TOKIMASA

*Technical Research Lab., Sumitomo Metal Industries, Ltd., Amagasaki, Japan*

## ABSTRACT

A detailed study was made on the effect of environmental air on the creep–fatigue fracture behavior of austenitic stainless steels and $2\frac{1}{4}$Cr–1Mo steel, based on the strain range partitioning concept. An accelerated testing procedure was found to obtain precisely a partitioned strain versus life correlation called the $\Delta\varepsilon_{ij}$–$N_{ij}$ relationship. The effect of environmental air on the $\Delta\varepsilon_{ij}$–$N_{ij}$ relationship was determined and the fracture morphology in creep–fatigue interaction was discussed.

Assuming that the $\Delta\varepsilon_{ij}$–$N_{ij}$ relationship is closely related to the crack growth properties, the ductility-normalized life equations in a perfect vacuum were derived and the partitioned strain versus crack growth rate equations were discussed.

## INTRODUCTION

Creep–fatigue interaction has recently become one of the most important aspects for the accurate fatigue design of high temperature structural components. Many studies have been made to evaluate the effect of creep–fatigue interaction on material strength at elevated temperatures. Most of the procedures for evaluating the creep effects on low cycle fatigue life have been developed by short-term accelerated testing. Since creep effects appear in prolonged service and since it is too time-consuming to simulate them at the same temperature and in the same loading environment as those in service, it is necessary to conduct accelerated creep–fatigue tests at higher temperatures or for shorter durations, and then to extrapolate the results obtained to lower temperatures or longer durations.

Among these accelerated creep–fatigue evaluating procedures, the strain range partitioning (SRP) creep–fatigue analysis proposed by Manson and co-workers [1–8] is considered the most promising to examine.

The author and his colleagues have been studying the creep–fatigue properties of materials for high temperature service such as those for nuclear and chemical power plants by the SRP approach since 1974 [9–18]. The effect of the strain wave form, thermal aging and environmental air on the low cycle fatigue behavior of austenitic stainless steels was successfully analyzed with the SRP concept, and it has been clarified that the creep–fatigue properties of many other structural materials can be well determined by this approach. In some cases, these properties can be successfully applied to evaluate the life of structural components [19, 20]. However, in the case when dynamic strain aging becomes severe, the SRP approach cannot be adequately applied and further studies have been required. Despite the fact that these limitations exist for the SRP approach, it still remains a hopeful method which is expected to be appropriately applied to the inelastic design of structural components.

In the present chapter appropriate experimental procedures and the results of a recent study on the environmental effects of creep–fatigue fracture behavior of austenitic stainless steels and $2\frac{1}{4}$Cr–1Mo steel are summarized.

## EXPERIMENTAL PROCEDURES

Manson and co-workers [1, 4] evaluated the $\Delta\varepsilon_{ij}$–$N_{ij}$ relationships by conducting tests with various strain wave forms. Among these should be noted the increasing ramp straining test (slow loading, fast unloading) and the decreasing ramp straining test (fast loading, slow unloading). In these tests, it is easy to separate $\Delta\varepsilon_{cp}$ or $\Delta\varepsilon_{pc}$ from $\Delta\varepsilon_{in}$, and it is possible to test a wider range of $\Delta\varepsilon_{cp}$ or $\Delta\varepsilon_{pc}$ values by changing the test temperature and the total strain range imposed. This is the essential difference from the strain-hold or stress-hold tests. In the case of the strain-hold test, it is very difficult to test for large values of $\Delta\varepsilon_{cp}$ or $\Delta\varepsilon_{pc}$, although it is very easy to partition the strain range. Both $\Delta\varepsilon_{cp}$ and $\Delta\varepsilon_{pc}$ have a tendency to saturate and take a small value, even if the strain-hold time is very long. In the case of the stress-hold test, it is impossible to predict the exact period of time which is necessary for the desired

TABLE 1

Practical procedures for PP, PC, CP and CC tests (*IJ* tests)

| Type of test | Strain wave form | Loading strain rate $\dot{\varepsilon}_1$ | Unloading strain rate $\dot{\varepsilon}_2$ | Type of inelastic strain | Partitioning of the strain range |
|---|---|---|---|---|---|
| PP | | Fast enough for creep not to occur ($\dot{\varepsilon}_1 = 10^{-3}$–$8 \times 10^{-3}$ s$^{-1}$) | $\dot{\varepsilon}_2 = \dot{\varepsilon}_1$ | $\Delta\varepsilon_{pp}$ | $\Delta\varepsilon_{in} = \Delta\varepsilon_{pp}$ (Fig. 1(a)) |
| PC | | The same as above | $\dot{\varepsilon}_2 \ll \dot{\varepsilon}_1$ ($\dot{\varepsilon}_2 < 10^{-4}$ s$^{-1}$) | $\Delta\varepsilon_{pp} + \Delta\varepsilon_{pc}$ | PP type of strain wave form was inserted during the test (Fig. 1(b)) |
| CP | | $\dot{\varepsilon}_1 < 10^{-4}$ s$^{-1}$ | Fast enough for creep not to occur ($\dot{\varepsilon}_1 = 10^{-3}$–$8 \times 10^{-3}$ s$^{-1}$) | $\Delta\varepsilon_{pp} + \Delta\varepsilon_{cp}$ | The same as above (Fig. 1(c)) |
| CC | | The same as above | $\dot{\varepsilon}_2 = \dot{\varepsilon}_1$ | $\Delta\varepsilon_{pp} + \Delta\varepsilon_{cc}$ | PC and CP types of strain wave forms were inserted during the test (Fig. 1(d)) |

value of $\Delta\varepsilon_{cp}$ or $\Delta\varepsilon_{pc}$ to be imposed in one cycle, although it is possible to test for a large value of $\Delta\varepsilon_{cp}$ or $\Delta\varepsilon_{pc}$. Since the amount of creep strain imposed in one cycle can be made larger in ramp straining tests, with the period of time for one cycle kept constant, it is possible to make the creep effect larger than the effect of high temperature oxidation, even if the tests are conducted in air.

Table 1 shows practical procedures for the four types of tests conducted by the author and his colleagues to obtain the $\Delta\varepsilon_{ij}$-$N_{ij}$ relationships. The half-cycle rapid load–unload method [4] was carried out for every specimen tested at a certain number of cycles during each test in order to determine the exact value of the partitioned inelastic strain range. Figure 1 shows details of the method to partition the inelastic strain range into $\Delta\varepsilon_{ij}$ in the $IJ$ test.

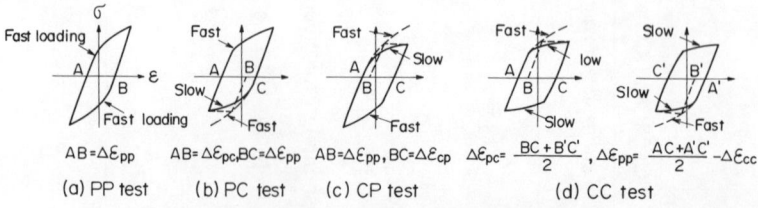

FIG. 1. Strain wave forms and hysteresis loops needed to analyze an inelastic strain range into the partitioned strain ranges $\Delta\varepsilon_{pp}$, $\Delta\varepsilon_{pc}$, $\Delta\varepsilon_{cp}$ and $\Delta\varepsilon_{cc}$ in (a) PP, (b) PC, (c) CP and (d) CC tests.

It was necessary to choose the loading strain rate $\dot\varepsilon_1$ and the unloading strain rate $\dot\varepsilon_2$ in order to perform these four types of tests successfully. Therefore, various combinations of $\dot\varepsilon_1$ and $\dot\varepsilon_2$ needed to be examined prior to the $IJ$ test and the hysteresis loop had to be partitioned. An example of the results of such a preliminary examination is shown in Fig. 2 for normalized and tempered $2\frac{1}{4}$Cr-1Mo steel, and it was found that valid combinations of $\dot\varepsilon_1$ and $\dot\varepsilon_2$ were $10^{-4}\,\text{s}^{-1}$ and $10^{-3}\,\text{s}^{-1}$ at 550 °C, and $10^{-4}\,\text{s}^{-1}$ and $8\times10^{-3}\,\text{s}^{-1}$ at 600 °C. When austenitic stainless steels were tested at temperatures from 550 to 600 °C, dynamic strain aging could appear and result in a negative $\Delta\varepsilon_{ij}$. In this case the test temperature had to be raised to avoid the appearance of dynamic strain aging. When high temperature oxidation had a large influence on the results, the tests needed to be made in a vacuum or in an inert gas.

The life relationships of $\Delta\varepsilon_{ij}$-$N_{ij}$ were derived by using the $\Delta\varepsilon_{pp}$-$N_{pp}$

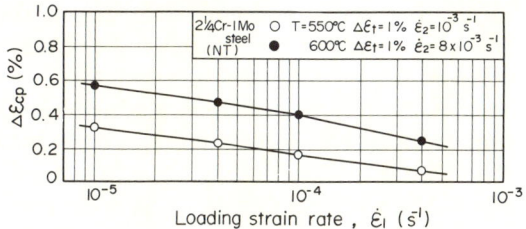

FIG. 2. Example of the observed relationship between $\Delta\varepsilon_{cp}$ and the loading strain rate $\dot{\varepsilon}_1$ in a CP test.

relationship, $\Delta\varepsilon_{pp}$ and $\Delta\varepsilon_{ij}$, $N_f$ in the *IJ* test, and the following linear damage rule:

$$\frac{1}{N_{pp}} + \frac{1}{N_{ij}} = \frac{1}{N_f} \tag{1}$$

## ENVIRONMENTAL EFFECT ON THE CREEP-FATIGUE PROPERTIES OF AUSTENITIC STAINLESS STEELS

The creep–fatigue properties of austenitic stainless steels have been studied and it has been found necessary to test at temperatures above 650 °C to avoid the marked effect of dynamic strain aging, and to show the prominent creep effect [9–12]. At temperatures higher than 650 °C in air, however, the effect of high temperature oxidation may become aggressive. Therefore, in order to obtain the accurate creep–fatigue properties of stainless steels, attention should be paid to the environmental effect on creep–fatigue interaction.

The test material selected for this purpose was SUS 304 austenitic stainless steel. The high temperature, low cycle fatigue tests (*IJ* tests) were conducted at 700 °C and 800 °C in both air and a vacuum test chamber, whose atmospheric pressure could be maintained below $6\cdot7 \times 10^{-4}$ Pa. Specimens were machined from forged and solution-treated bars of diameter 28 mm and length 400 mm, whose solution treatment involved annealing at 1100 °C for 30 min and subsequent water cooling. The specimen diameter was 10 mm and its gauge length was 25 mm. All specimens were carefully polished with #400 emery paper to remove any circumferential machining striations. An MTS

25 ton electro-hydraulic servo-controlled test system was used with an induction heater and a vacuum test chamber. The strain wave forms were programmed with a digital function generator.

Figure 3 shows the $\Delta\varepsilon_{ij}$-$N_{ij}$ relationships obtained from $IJ$ tests in air and in a vacuum. As shown in Fig. 3(a) the $\Delta\varepsilon_{pp}$-$N_{pp}$ relationships in air and vacuum take significantly differed forms. It was noted, however, that temperatures of 700°C and 800°C had little effect on this relationship. The fatigue life in a vacuum was about four times as long as that in air, indicating that the $\Delta\varepsilon_{pp}$-$N_{pp}$ relationship is a property very sensitive to the environmental air.

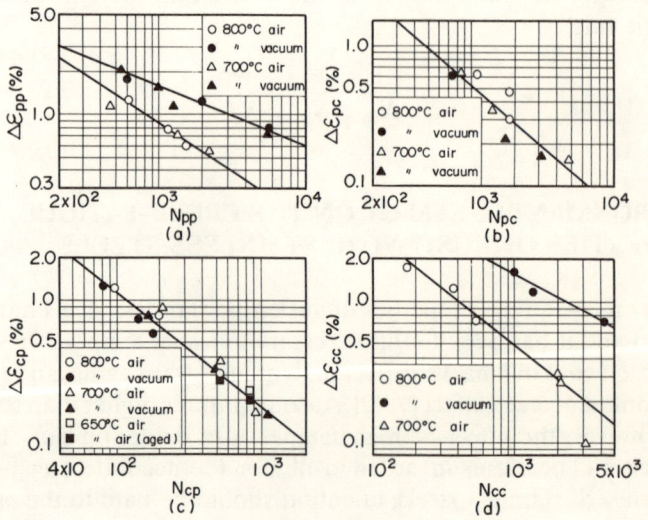

FIG. 3. Partitioned strain versus life relationships of SUS 304 stainless steel at 700 and 800°C in air and a vacuum.

The $\Delta\varepsilon_{pc}$-$N_{pc}$ relationship shown in Fig. 3(b) tends to fall on a line in a log–log plot, suggesting that the $\Delta\varepsilon_{pc}$-$N_{pc}$ relationship was insensitive to the test temperature and environment.

All the data for the $\Delta\varepsilon_{cp}$-$N_{cp}$ relationship shown in Fig. 3(c) also tend to fall on a line in a log–log plot with little scatter. In this figure, the data obtained at 650°C were also plotted, and it can be said from this that the $\Delta\varepsilon_{cp}$-$N_{cp}$ relationship was not dependent on the test temperature in the range from 650 to 800°C, and that there was no environmental effect on this relationship.

Figure 3(d) indicates the marked environmental effect on the $\Delta\varepsilon_{cc}$-$N_{cc}$ relationship. The fatigue life in a vacuum was about five times as long as that in air.

In general, fatigue properties are very sensitive to the environment when the deformation mode in the loading and unloading period is the same (the $\Delta\varepsilon_{pp}$-$N_{pp}$ and $\Delta\varepsilon_{cc}$-$N_{cc}$ relationships). However, fatigue properties seem to be insensitive to the environment when the deformation mode in the loading and unloading period is not the same (the $\Delta\varepsilon_{pc}$-$N_{pc}$ and $\Delta\varepsilon_{cp}$-$N_{cp}$ relationships).

The fact that the $\Delta\varepsilon_{pp}$-$N_{pp}$ and $\Delta\varepsilon_{cc}$-$N_{cc}$ relationships were sensitive to the environment and that the fatigue life in vacuum was four to five times longer than that in air can be qualitatively explained by the crack growth behavior. Transgranular or intergranular cracks were initiated on the specimen surface, and one of the cracks grew to a main crack in a PP or CC test. When such a reversible cyclic inelastic strain as $\Delta\varepsilon_{pp}$ or $\Delta\varepsilon_{cc}$ was imposed on the specimen, the specimen surface, where the stress state was biaxial, suffered from larger fatigue damage than the interior of the specimen, where the stress state was microscopically triaxial. Therefore, the cracks tended to be initiated at the specimen surface, and the main crack was exposed to the effect of the environment in both crack initiation and growth. Figure 4 shows an example of the fracture surface of a PP-tested specimen. The main crack was initiated at the specimen surface and grew to a final failure.

The reason that the $\Delta\varepsilon_{pc}$-$N_{pc}$ and $\Delta\varepsilon_{cp}$-$N_{cp}$ relationships were insensitive to the environment can also be explained by the crack initiation and growth behavior. In the case of $\Delta\varepsilon_{pc}$-type inelastic strain cycling, an accumulation of tensile plastic strain led the specimen to a tensile fracture, and this process is insensitive to the environment. In the case of $\Delta\varepsilon_{cp}$ type inelastic straining the fatigue life is governed by the initiation and coalescence or connection of grain-boundary cracks inside the specimen, and this process is also insensitive to the environment. Figure 5 shows an example of the fracture surface of a CP-tested specimen, showing typical intergranular crack growth.

Figure 6 shows the microphotographs obtained at the center of specimens near the fracture surface. No origin of the crack can be seen at the center of both the PP- and the CC-tested specimens. In the PC-tested specimen, larger tensile plastic deformation can be seen and the inclusions were elongated in the axial direction. Many dimples were observed with a scanning electron micrograph on the fracture surface. In the CP-tested specimen, a large number of typical intergranular cracks are visible. Therefore, it is clear that the specimen failure was

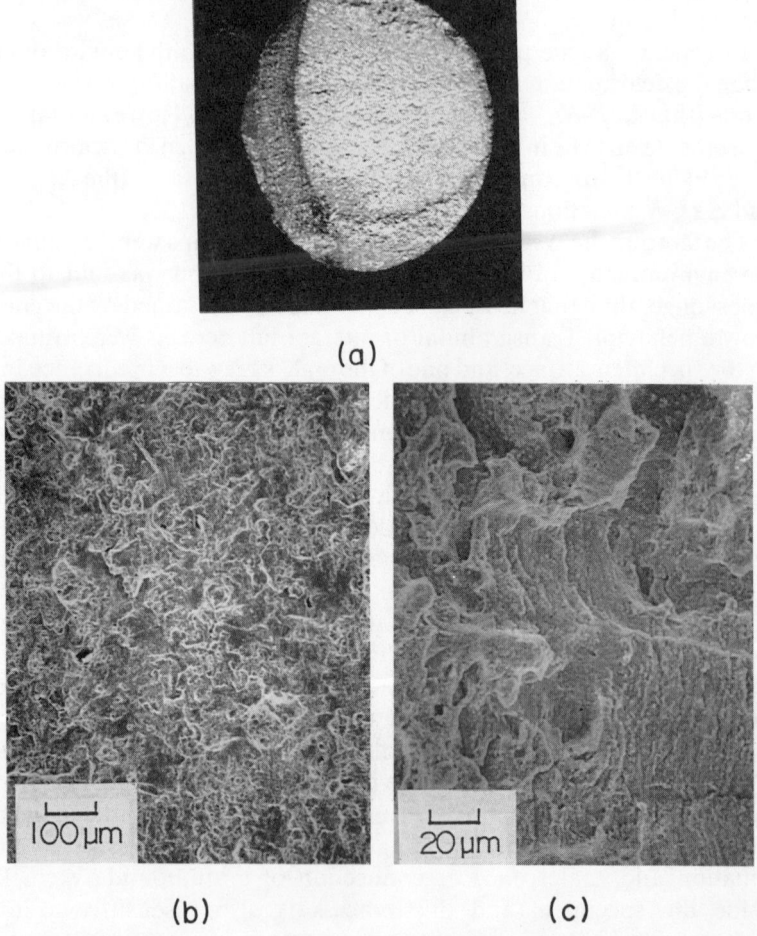

(a)

(b)                                        (c)

FIG. 4. Examples of the fracture surface of a PP-tested specimen: (a) macroscopic view; (b) scanning electron micrograph; (c) magnified view of the left-hand portion of (b).

caused by the coalescence and/or connection of grain boundary cracks inside the specimen in the case of CP-type strain cycling.

Most of the investigations made on the environmental effects of high temperature fatigue life have reported the significance of such effects. This is not surprising in the light of the results just presented, since those reported results were obtained under such a cyclic inelastic strain as a

$\Delta\varepsilon_{pp}$ or $\Delta\varepsilon_{pp} + \Delta\varepsilon_{cc}$ type. From the same standpoint, the study of the environmental effects on crack growth behavior by Solomon and Coffin [21] is meaningful in such a case that a $\Delta\varepsilon_{pp}$ or $\Delta\varepsilon_{pp} + \Delta\varepsilon_{cc}$ type of inelastic strain was cycled.

Another interesting finding is that the $\Delta\varepsilon_{pp}$-$N_{pp}$ and $\Delta\varepsilon_{cc}$-$N_{cc}$ relationships took the same form in a vacuum, and they could be expressed by Manson's type of equation as shown in Fig. 7.

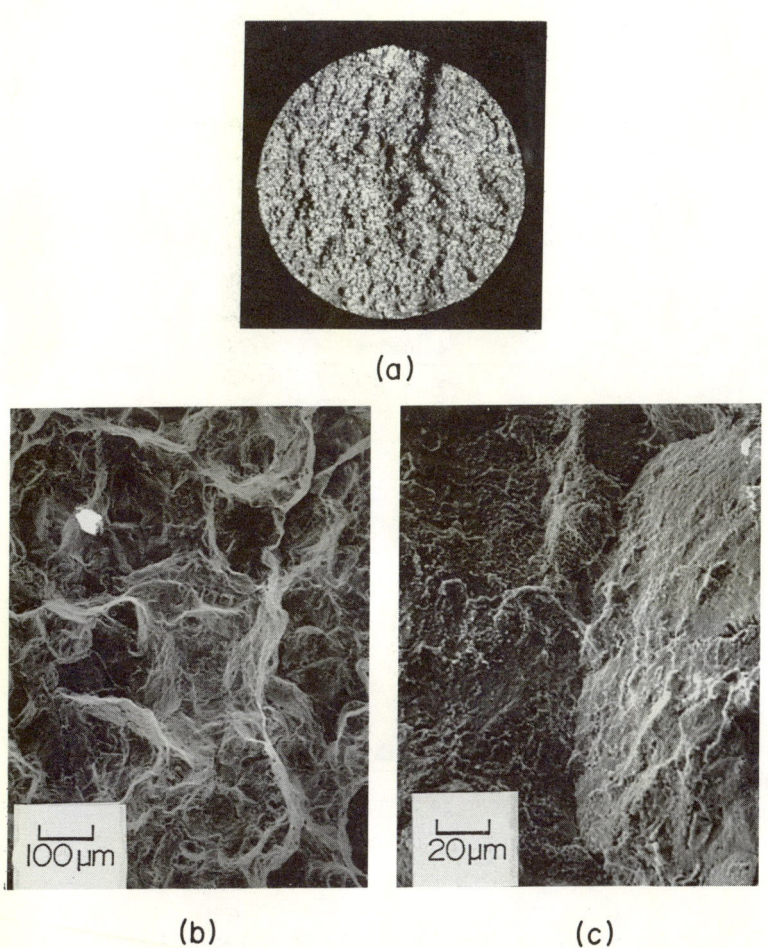

(a)

(b)                              (c)

FIG. 5. Examples of the fracture surface of a CP-tested specimen: (a) macroscopic view; (b) scanning electron micrograph; (c) magnified view of the left-hand portion of (b).

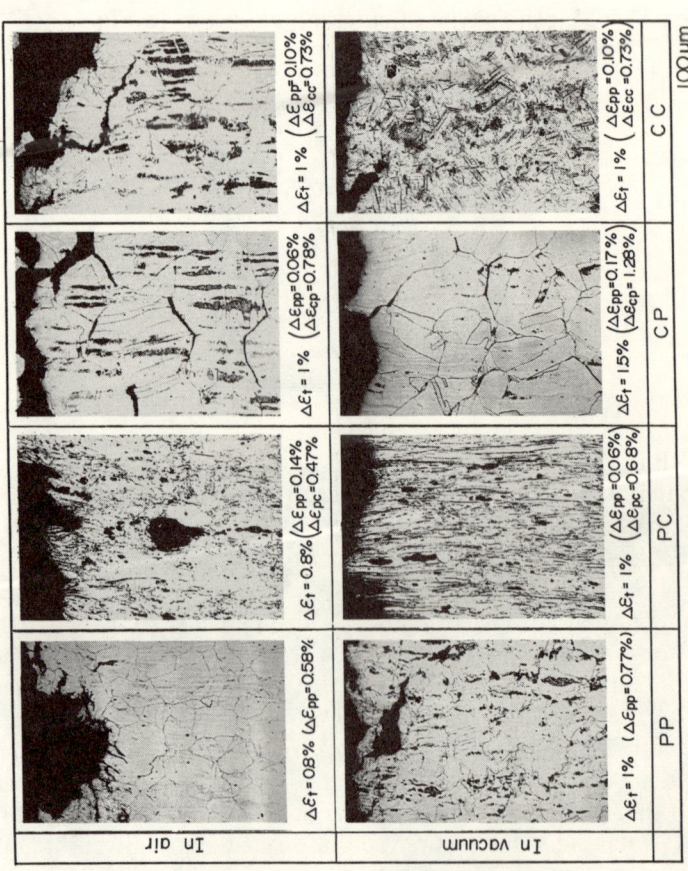

FIG. 6. Microphotographs obtained at the center of specimens near the fracture surface ($T = 800°C$).

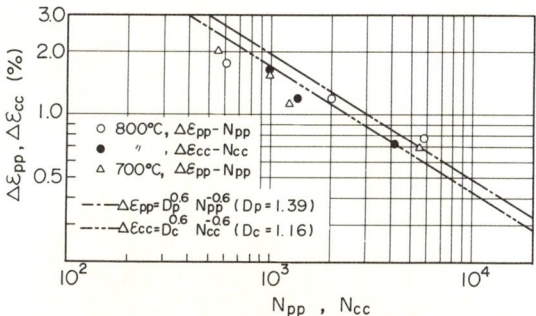

FIG. 7. Comparison between the experimental and calculated results of the $\Delta\varepsilon_{pp}$-$N_{pp}$ and $\Delta\varepsilon_{cc}$-$N_{cc}$ relationships in a vacuum.

## ENVIRONMENTAL EFFECT ON THE CREEP–FATIGUE PROPERTIES OF $2\frac{1}{4}$Cr–1Mo STEEL

The SRP life relationships were also determined for the annealed and the normalized and tempered $2\frac{1}{4}$Cr–1Mo steels at 550 °C in both air and vacuum, and the effects of the environmental air and heat treatment on the SRP life relationships of $2\frac{1}{4}$Cr–1Mo steel were examined [16]. The test materials were forged to a round bar 23 mm in diameter and then isothermally annealed or normalized and tempered.

The $\Delta\varepsilon_{ij}$-$N_{ij}$ relationships in air and in vacuum are shown in Fig. 8 and Fig. 9 respectively, and the effects of environmental air on these relationships are summarized in Fig. 10. As can be seen in these figures, all the $\Delta\varepsilon_{ij}$-$N_{ij}$ properties tended to be sensitive to the environment in the strain range lower than a given value of $\Delta\varepsilon_{ij}$. In the annealed material, the tendency was for $N_{pc}$ to be less than or equal to $N_{cp}$. This can be attributed to the environmental effect, because the vacuum data for both materials show that $N_{cp}$ tended to be less than $N_{pc}$, independent of the value of the inelastic strain range. It was also found that all the $\Delta\varepsilon_{cc}$-$N_{cc}$ data under this vacuum condition gave shorter lives than those predicted by Manson's equation, whereas Manson's equation satisfactorily predicted $N_{pp}$ for the vacuum condition. This fact suggests that the vacuum condition used (less than $6\cdot7 \times 10^{-4}$ Pa) was not ideal for obtaining the SRP life relationships in a vacuum for $2\frac{1}{4}$Cr–1Mo steel.

Figure 11 shows a proposed model of the environmental effect on the SRP life relationships, which is based on the results just described and

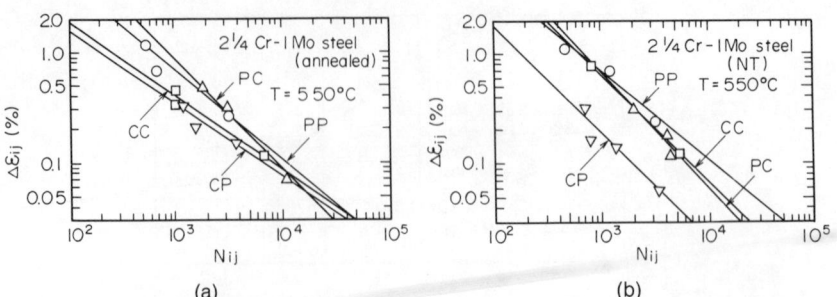

FIG. 8. $\Delta\varepsilon_{ij}$–$N_{ij}$ relationships for $2\frac{1}{4}$Cr–1Mo steel at 550°C in air: (a) annealed material; (b) normalized and tempered material.

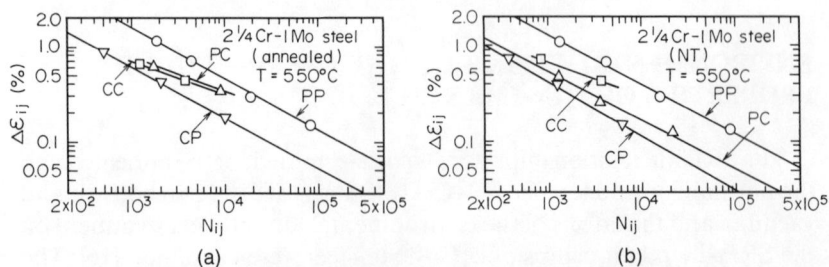

FIG. 9. $\Delta\varepsilon_{ij}$–$N_{ij}$ relationships for $2\frac{1}{4}$Cr–1Mo steel at 550°C in a vacuum: (a) annealed material; (b) normalized and tempered material.

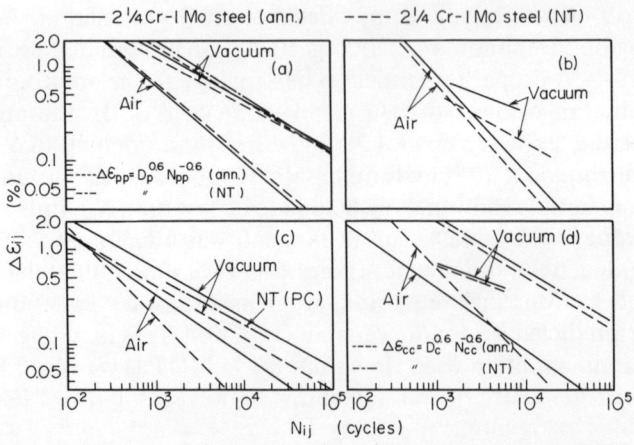

FIG. 10. Effects of the environment and heat treatment on the $\Delta\varepsilon_{ij}$–$N_{ij}$ relationships of $2\frac{1}{4}$Cr–1Mo steel: (a) PP; (b) PC; (c) CP; (d) CC.

the facts reported in the literature [22–24]. Using this model, it was possible to obtain the SRP life relationship in a perfect vacuum based on both air data and the non-ideal vacuum data, where it was assumed that all the SRP life relationships obtained in various environments would intersect at a point O depending on the type of inelastic strain range. It was also assumed that the SRP life relationships in a perfect vacuum would be given by the following equation:

$$\Delta \varepsilon_{ij} = \alpha_{ij} D_i N_{ij}^{-0.5} \qquad i,j = \text{p, c} \qquad (2)$$

This equation is analogous to Coffin's equation, and it was adopted in place of Manson's equation, because the vacuum used was not ideal and

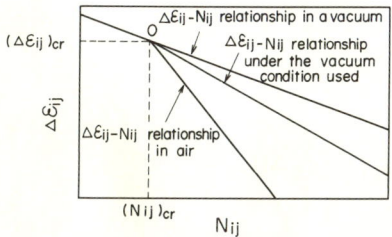

FIG. 11. Proposed model of the environmental effect on the SRP life relationship.

the latter equation cannot describe the effect of the environment on $N_{pp}$ and $N_{cc}$ in this non-ideal vacuum. Then

$$\alpha_{pp} = \alpha_{cc} = 0.5 \qquad (3)$$

Figure 12 shows the results of a comparision between Eqn. (2) and the experimental data for the normalized and tempered material. Figure 13 shows the SRP life relationships in a perfect vacuum for the annealed and the normalized and tempered material, which clearly indicates that the effect of heat treatment on the $\Delta \varepsilon_{pp}$-$N_{pp}$ and $\Delta \varepsilon_{cc}$-$N_{cc}$ properties was not marked, whereas the $\Delta \varepsilon_{pc}$-$N_{pc}$ and $\Delta \varepsilon_{cp}$-$N_{cp}$ properties of the annealed material were superior to those of the normalized and tempered material.

The fracture surfaces obtained from $IJ$ tests in a vacuum were examined with a scanning electron microscope, the results being shown in Fig. 14. The fracture surfaces of the CP-tested specimens were quite different from those of the PP-, PC- and CC-tested specimens and seem to indicate that crack initiation and growth from inside the specimen

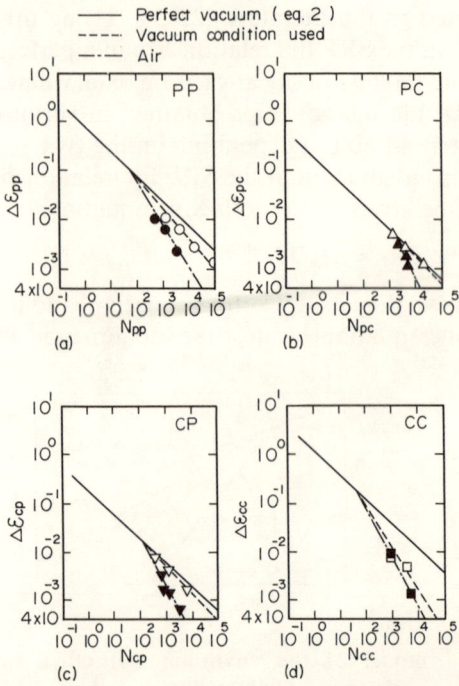

FIG. 12. Comparison of Eqn. (2) with experimental data for normalized and tempered $2\frac{1}{4}$Cr–1Mo steel: (a) $\Delta\varepsilon_{pp}$–$N_{pp}$ relationship; (b) $\Delta\varepsilon_{pc}$–$N_{pc}$ relationship; (c) $\Delta\varepsilon_{cp}$–$N_{cp}$ relationship; (d) $\Delta\varepsilon_{cc}$–$N_{cc}$ relationship.

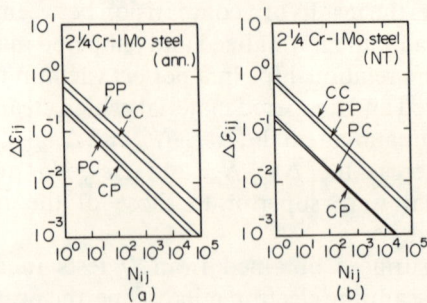

FIG. 13. Summary of the SRP life relationships in a perfect vacuum for (a) annealed and (b) normalized and tempered $2\frac{1}{4}$Cr–1Mo steels.

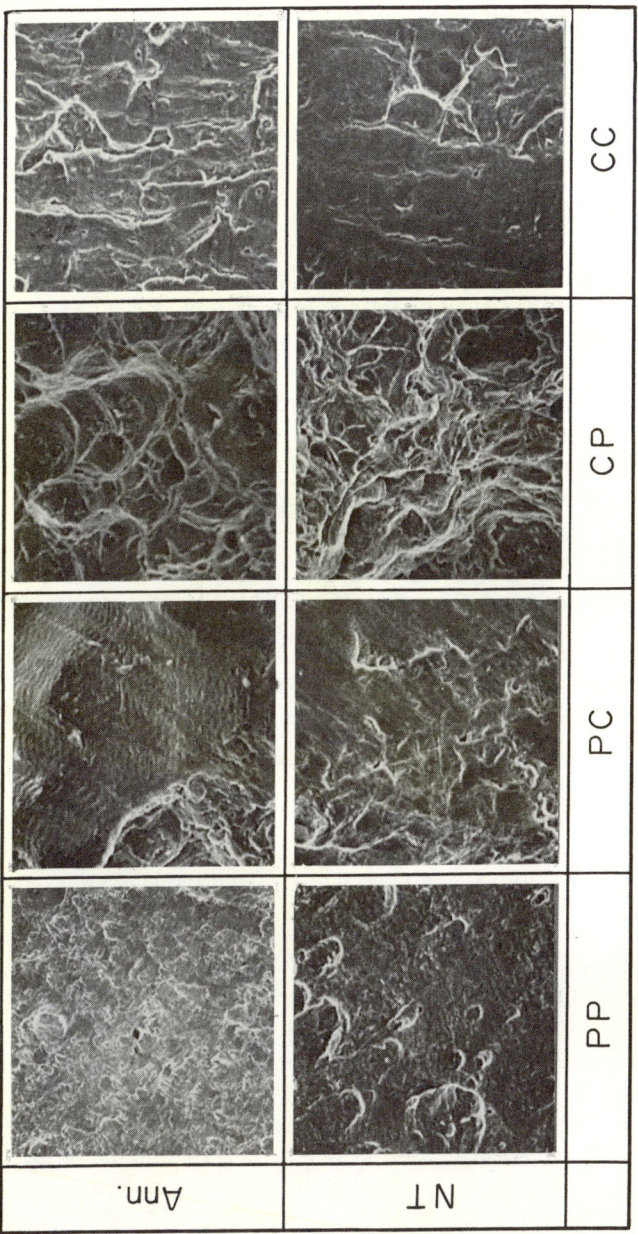

FIG. 14. Scanning electron micrographs of the fracture surfaces obtained from $IJ$ tests in a vacuum for $2\frac{1}{4}$Cr–1Mo steel.

occurred and that it was caused by void formation and coalescence. The fracture surfaces of the PP-, PC- and CC-tested specimens are typical of those that can be obtained by a normal fatigue crack growth test, suggesting that the effect of the environment must be taken into account in the vacuum condition used for these tests.

## FRACTURE MORPHOLOGY IN CREEP–FATIGUE

A fractographic study was conducted on austenitic stainless steels fractured in the five basic creep–fatigue tests, namely PP, PC, CP, CC, and DC tests. The type of inelastic strain cycled to the specimen in the DC test was $N_i \, \Delta\varepsilon_{pp} + \delta_c$, where $N_i$ is the number of cycles of $\Delta\varepsilon_{pp}$ in one cycle and $\delta_c$ is the creep ratchet strain, also in one cycle [13]. The fracture surface was observed with a scanning electron microscope and the crack growth behavior was examined in the longitudinal cross-section of the specimen with an optical microscope.

TABLE 2

Features of the fracture appearance under the basic strain conditions for creep–fatigue interaction

| Type of cycled inelastic strain | Fracture appearance | | Main cause of fracture |
|---|---|---|---|
| | Surface | Inside | |
| $\Delta\varepsilon_{pp}$ | TGC with fatigue striations | TGC with fatigue striations | TGCG from the surface to inside |
| $\Delta\varepsilon_{pp} + \Delta\varepsilon_{pc}$ | TGC with fatigue striations | Elongated dimples or voids | Void formation |
| $\Delta\varepsilon_{pp} + \Delta\varepsilon_{cp}$ | T and IGC with fatigue striations | IGC | Connection or coalescence of many IGC inside the specimen |
| $\Delta\varepsilon_{pp} + \Delta\varepsilon_{cc}$ | IGC | IGC | IGCG from the surface to inside |
| $N_i \, \Delta\varepsilon_{pp} + \delta_c$ | TGC due to $\Delta\varepsilon_{pp}$ IGC due to $\delta_c$ | | Intermittent TGCG and IGCG from the surface to inside |

Notes: TGC, transgranular crack; IGC, intergranular crack; TGCG, transgranular crack growth; IGCG, intergranular crack growth.

TABLE 3

Predicted effect of environment, temperature and imposed stress or strain level on the fracture appearance under the basic strain conditions for creep-fatigue interaction

| Type of strain | Effect of environment | Effect of elevating temperature (without environmental effects) | Effect of increasing strain level (without environmental effects) |
|---|---|---|---|
| $\Delta\varepsilon_{pp}$ | Crack growth rate acceleration TGCG → IGCG Striation formation | None | Fatigue striation → IGC |
| $\Delta\varepsilon_{pc}$ | None | None | Void formation → IGC |
| $\Delta\varepsilon_{cp}$ | None | Transgranular connection or coalescence of IGC → Intergranular one Cavity formation at grain boundaries | Round-type IGC → Wedge-type IGC |
| $\Delta\varepsilon_{cc}$ | Acceleration of IGCG Striation formation | Cavity formation at grain boundaries | Round-type IGC → Wedge-type IGC |
| $\delta_c$ | Acceleration or deceleration of IGCG | Cavity formation at grain boundaries | Round-type IGC → Wedge-type IGC |

The features of the fracture appearance under the basic strain conditions for creep–fatigue interaction are summarized in Table 2, indicating that the fracture morphologies were closely related to the type of test, i.e. the type of inelastic strain cycled to the specimen.

When the $\Delta\varepsilon_{pp}$ type of strain was cycled, transgranular crack initiation and its growth at the specimen surface were typical. When the $\Delta\varepsilon_{pc}$ type of strain was cycled, a dimple formation occurred inside the specimen and led to a tensile-like fracture. When the $\Delta\varepsilon_{cp}$ type of strain was cycled, the initiation of many cracks at the grain boundaries and their intergranular or transgranular coalescence were typical. When a $\Delta\varepsilon_{cc}$ type of strain or $\delta_c$ type of strain was cycled, intergranular crack initiation and its growth in the specimen were fundamental features of the fracture surface.

Based on these results, it was possible to deduce the effects of environment, temperature and stress or strain condition on each fracture morphology. Table 3 summarizes the predicted effects of environment, temperature and imposed stress or strain level on the fracture appearance under the basic strain condition. The fundamental mechanism for nucleation of voids at the grain boundary seems to be the probable explanation for the difference between the fracture morphologies.

## RELATIONSHIP BETWEEN THE HIGH TEMPERATURE CREEP–FATIGUE PROPERTIES AND THE TENSILE AND CREEP DUCTILITIES

It is very useful to clarify the relationship between the high temperature creep–fatigue properties and the tensile and creep ductilities, not only for the design of structural components, but also for the development of the materials. In Table 4 are shown ductility-normalized strain range partitioning (DN-SRP) life relationships proposed by NASA researchers [2, 8]. These equations were obtained from test data in air, and so they are not always the general equations since the environmental effect is present in air and is considered to be independent of the material ductility as already described.

From this information, it seemed to be important to clarify the DN-SRP life relationships in a perfect vacuum and to understand the environmental effect in these life relationships. The author has tried to obtain the DN-SRP life relationships with several assumptions [17].

## TABLE 4
DN-SRP life relationships reported in the literature [2, 8]

| *Author* | *DN–SRP life relationship* | |
|---|---|---|
| Manson (1973) [2] | $\Delta\varepsilon_{pp} = 0{\cdot}75D_pN_{pp}^{-0\cdot6}$ | $\Delta\varepsilon_{pc} = 1{\cdot}25D_pN_{pc}^{-0\cdot8}$ |
| | $\Delta\varepsilon_{cp} = 0{\cdot}25D_cN_{cp}^{-0\cdot8}$ | $\Delta\varepsilon_{cc} = 0{\cdot}75D_cN_{cc}^{-0\cdot8}$ |
| Halford *et al.* (1977) [8] | $\Delta\varepsilon_{pp} = 0{\cdot}5D_pN_{pp}^{-0\cdot6}$ | $\Delta\varepsilon_{pc} = 0{\cdot}25D_pN_{pc}^{-0\cdot6}$ |
| | $\Delta\varepsilon_{cp} = 0{\cdot}2D_c^{0\cdot6}N_{cp}^{-0\cdot6}$ | (transgranular crack) |
| | $\Delta\varepsilon_{cp} = 0{\cdot}1D_c^{0\cdot6}N_{cp}^{-0\cdot6}$ | (intergranular crack) |
| | $\Delta\varepsilon_{cc} = 0{\cdot}25D_c^{0\cdot6}N_{cc}^{-0\cdot6}$ | |

$D_p$, tensile ductility; $D_c$, creep ductility.

The following were assumed in the derivation of the DN-SRP life relationships in a perfect vacuum.

(1) The $\Delta\varepsilon_{pp}$-$N_{pp}$ and $\Delta\varepsilon_{cc}$-$N_{cc}$ relationships can be represented by Coffin's equation as

$$\Delta\varepsilon_{ii} = 0{\cdot}5D_iN_{ii}^{-0\cdot5} \qquad (4)$$

(2) When $\Delta\varepsilon_{ij}$ is large, the crack initiation life is small enough compared with the crack propagation life and can be neglected.

(3) The same equations can be applied to describe the crack propagation behavior when the tensile deformation mode in cyclic inelastic strain ($\Delta\varepsilon_{ij}$) is the same.

(4) The usual relationship between the crack propagation rate and plastic strain range [25] is assumed to hold for the relationships between the crack propagation rate and inelastic strain range ($\Delta\varepsilon_{ij}$).

Under these assumptions it was possible to derive the same DN-SRP life relationships as Eqn. (2) [17], i.e.

$$\Delta\varepsilon_{ij} = \alpha_{ij}D_iN_{ij}^{-0\cdot5} \qquad (2)$$

The DN-SRP life relationships could then be determined by obtaining $\alpha_{ij}$.

In Table 5 the values of $\alpha_{ij}$ obtained for the annealed and the normalized and tempered $2\frac{1}{4}$Cr-1Mo steels are listed. The maximum difference between the values obtained for two materials was a factor of 2, and here the mean values were used to express the DN-SRP life relationships.

TABLE 5
$D_p$, $D_c$, $\alpha_{pc}$ and $\alpha_{cp}$ for $2\frac{1}{4}$Cr–1Mo steels at 550 °C

| Material | $D_p$ | $D_c$ | $\alpha_{pc}$ | $\alpha_{cp}$ |
|----------|-------|-------|---------------|---------------|
| Annealed | 2·01  | 1·86  | 0·167         | 0·147         |
| NT       | 1·54  | 2·36  | 0·123         | 0·074         |
| (Averaged) | —   | —     | (0·145)       | (0·111)       |

Finally, the following DN-SRP life relationships were determined:

$$\Delta\varepsilon_{pp} = 0\cdot5 D_p N_{pp}^{-0.5}$$
$$\Delta\varepsilon_{pc} = 0\cdot145 D_p N_{pc}^{-0.5}$$
$$\Delta\varepsilon_{cp} = 0\cdot111 D_c N_{cp}^{-0.5} \tag{5}$$
$$\Delta\varepsilon_{cc} = 0\cdot5 D_c N_{cc}^{-0.5}$$

Figure 15 shows the applicability of Eqns. (5) to SUS 304 steels. The $D_c$ value was determined from the creep rupture data obtained within 500 h at 700 °C as 0·72. The experimental data agree well with the DN-SRP life relationships obtained, except for one or two data points in the $\Delta\varepsilon_{pc}$–$N_{pc}$ and $\Delta\varepsilon_{cp}$–$N_{cp}$ relationships.

By using Eqns. (5) and the DN-SRP life relationships proposed by Halford et al. [8] that are shown in Table 4, the effect of environmental air could be quantitatively given. If the effect of environmental air could be given as the ratio of the life in air to that in vacuum, $k_{ij} = (N_{ij})_{air}/(N_{ij})_{vacuum}$, and $k_{pp}$, $k_{pc}$, $k_{cp}$ (transgranular), $k_{cp}$ (intergranular) and $k_{cc}$ would be $1\cdot26\,(\Delta\varepsilon_{pp}/D_p)^{1/3}$, $4\cdot72\,(\Delta\varepsilon_{pc}/D_p)^{1/3}$, $5\cdot55\,\Delta\varepsilon_{cp}^{1/3}/D_c$, $1\cdot75\,\Delta\varepsilon_{cp}^{1/3}/D_c$ and $0\cdot391\,\Delta\varepsilon_{cc}^{1/3}/D_c$, respectively. It was found that the degree of environmental effect depended on $\Delta\varepsilon_{ij}$ and the ductility.

## CREEP–FATIGUE CRACK GROWTH RATE EQUATIONS

The SRP analysis of creep–fatigue cracks was conducted for the normalized and tempered $2\frac{1}{4}$Cr–1Mo steel at 550 °C, and for the SUS 304 stainless steel at 700 °C, assuming that creep–fatigue life relationships could be obtained by integrating the creep–fatigue crack growth rate equations [18].

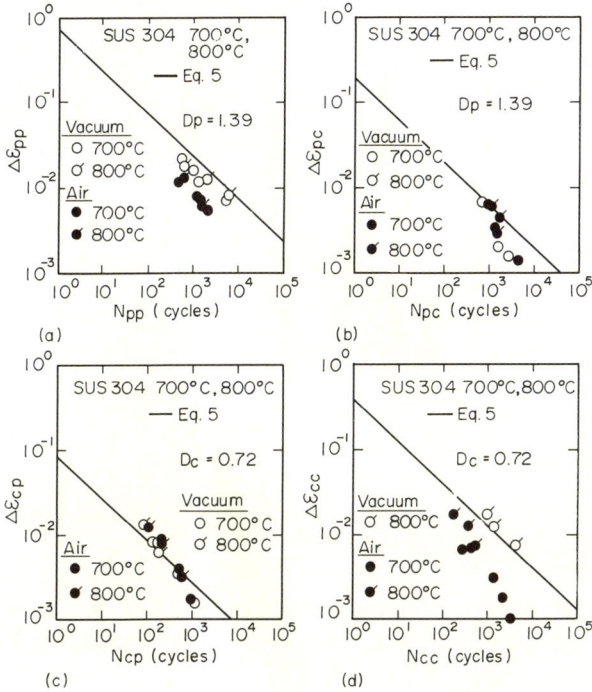

FIG. 15. DN-SRP life relationships (Eqn. (5)) and experimental data obtained at 700 and 800 °C for SUS 304 stainless steel: (a) $\Delta\varepsilon_{pp}$-$N_{pp}$ relationship; (b) $\Delta\varepsilon_{pc}$-$N_{pc}$ relationship; (c) $\Delta\varepsilon_{cp}$-$N_{cp}$ relationship; (d) $\Delta\varepsilon_{cc}$-$N_{cc}$ relationship.

Figure 16 shows a schematic illustration of a proposed model of the relationship between the $\Delta\varepsilon_{ij}$-$N_{ij}$ properties and crack growth properties, which can be described by the following equations.

(1) The $\Delta\varepsilon_{ij}$-$N_{ij}$ relationships in a vacuum:

$$\Delta\varepsilon_{ij} = \alpha_{ij}D_iN_{ij}^{-0.5} \qquad (2)$$

(2) The $\Delta\varepsilon_{ij}$-$N_{ij}$ relationships in air:

$$\Delta\varepsilon_{ij} = A_{ij}N_{ij}^{-m_{ij}} \qquad \text{for } \Delta\varepsilon_{ij} \leqslant (\Delta\varepsilon_{ij})_{cr}$$
$$\Delta\varepsilon_{ij} = \alpha_{ij}D_iN_{ij}^{-0.5} \qquad \text{for } \Delta\varepsilon_{ij} > (\Delta\varepsilon_{ij})_{cr} \qquad (6)$$

(3) The $(da/dN)_{ij}$ versus $\Delta\varepsilon_{ij}$ relationships in a vacuum:

$$1/a\,(da/dN)_{ij} = \ln{(a_f/a_0)_{ij}}(\Delta\varepsilon_{ij}/\alpha_{ij}D_i)^2 \qquad (7)$$

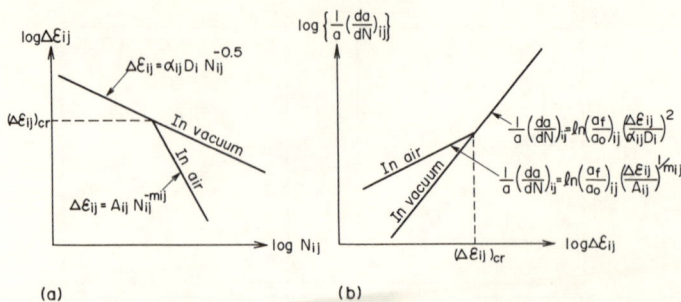

FIG. 16. Proposed model of the relationship between the partitioned creep-fatigue life and crack growth rate: (a) $\Delta\varepsilon_{ij}$ versus $N_{ij}$ relationship; (b) $(da/dN)_{ij}$ versus $\Delta\varepsilon_{ij}$ relationship.

(4) The $(da/dN)_{ij}$ versus $\Delta\varepsilon_{ij}$ relationships in air:

$$1/a\,(da/dN)_{ij} = \ln(a_f/a_0)_{ij}(\Delta\varepsilon_{ij}/A_{ij})1/m_{ij} \quad \text{for } \Delta\varepsilon_{ij} \leqslant (\Delta\varepsilon_{ij})_{cr}$$
$$1/a\,(da/dN)_{ij} = \ln(a_f/a_0)_{ij}(\Delta\varepsilon_{ij}/\alpha_{ij}D_i)^2 \quad \text{for } \Delta\varepsilon_{ij} > (\Delta\varepsilon_{ij})_{cr} \tag{8}$$

where $A_{ij}$ and $m_{ij}$ are constant depending on the material and temperature, and $(a_f/a_0)_{ij}$ is the ratio of the final to initial crack size from $\Delta\varepsilon_{ij}$ straining.

From these equations, either the $\Delta\varepsilon_{ij}$ versus $N_{ij}$ relationships or the $(da/dN)_{ij}$ versus $\Delta\varepsilon_{ij}$ relationships can be determined when $(a_f/a_0)_{ij}$ are given. Conversely, $(a_f/a_0)_{ij}$ can be determined when both the $\Delta\varepsilon_{ij}$ versus $N_{ij}$ relationships and the $(da/dN)_{ij}$ versus $\Delta\varepsilon_{ij}$ relationships are given.

In order to determine the crack growth properties, creep–fatigue crack growth tests were conducted with a centrally notched plate specimen (20·6 mm width × 5 mm thickness, initial artificial crack length = 1·0 mm, gage length = 25 mm) on $2\frac{1}{4}$Cr-1Mo and SUS 304 steels under the $IJ$ test conditions. Based on the $IJ$ test results, the $(da/dN)_{ij}$ versus $\Delta\varepsilon_{ij}$ relationships were calculated by the following procedures. It was assumed that the crack growth increment $\Delta a$ by $\Delta\varepsilon_{in}$ ($= \Sigma\Delta\varepsilon_{ij}$) was equal to $\Sigma(\Delta a)_{ij}$ (linear summation rule). Under this assumption, the crack growth rate $da/dN$ is given by $\Sigma(da/dN)_{ij}$, i.e.

$$da/dN = \Sigma(da/dN)_{ij} \tag{9}$$

First, with the PP test results at a given $\Delta\varepsilon_{pp}$ level, the value of $1/a\,(da/dN)_{pp}$ corresponding to the $\Delta\varepsilon_{pp}$ value can be determined. Then, with the $IJ$ test result for a given $\Delta\varepsilon_{in}$ ($= \Delta\varepsilon_{pp} + \Delta\varepsilon_{ij}$) level, the value of

$1/a(da/dN)_{ij}$ corresponding to the $\Delta\varepsilon_{ij}$ value can be obtained using Eqn. (9). By substituting the experimental values $(\Delta\varepsilon_{ij})_0$ and $[1/a(da/dN)_{ij}]_0$ into Eqn. (7) or Eqn. (8), the value of ln $(a_f/a_0)_{ij}$ can be obtained when the values $\alpha_{ij}$, $D_i$, $A_{ij}$ and $m_{ij}$ are known.

TABLE 6

Creep-fatigue properties of normalized and tempered $2\frac{1}{4}$Cr-1Mo steel and SUS 304 steel

| Material | $ij$ | $a_{ij}$ | $D_i$ | $A_{ij}$ | $1/m_{ij}$ | $(\Delta\varepsilon_{ij})_{cr}$ |
|---|---|---|---|---|---|---|
| $2\frac{1}{4}$Cr-1Mo (NT) | pp | 0·5 | 1·54 | 2·02 | 1·23 | 0·164 |
| $T = 550\,°C$ | pc | 0·145 | | 11·4 | 0·935 | $7·07 \times 10^{-3}$ |
| | cp | 0·111 | 2·36 | 1·49 | 1·04 | 0·040 4 |
| | cc | 0·5 | | 6·03 | 1·01 | 0·221 |
| SU 304 | pp | 0·5 | 1·39 | 1·03 | 1·44 | 0·250 |
| $T = 700\,°C$ | pc | 0·145 | | 2·33 | 1·13 | $8·21 \times 10^{-3}$ |
| | cp | 0·111 | 0·72 | 0·444 | 1·24 | $4·76 \times 10^{-3}$ |
| | cc | 0·5 | | 3·04 | 1·02 | 0·039 8 |

The creep-fatigue properties of the normalized and tempered $2\frac{1}{4}$Cr-1Mo steel and SUS 304 stainless steel are shown in Table 6. The $da/dN$ versus crack length relationships obtained by $IJ$ tests in air are shown in Fig. 17. Based on these results, the values of the ln $(a_f/a_0)_{ij}$ and $(da/dN)_{ij}$ versus $\Delta\varepsilon_{ij}$ relationships were determined. The $(da/dN)_{ij}$ versus $\Delta\varepsilon_{ij}$ relationship can be expressed as follows:

$$(da/dN)_{ij} = B_{ij}(\Delta\varepsilon_{ij})^2 a \qquad \text{in a vacuum} \qquad (10)$$

or

$$(da/dN)_{ij} = B'_{ij}\,\Delta\varepsilon_{ij}^{1/m_{ij}} a \qquad (11)$$

where $B_{ij} = \ln(a_f/a_0)_{ij}/(\alpha_{ij}D_i)^2$ and $B'_{ij} = \ln(a_f/a_0)_{ij}/A_{ij}^{1/m_{ij}}$.

The values of ln $(a_f/a_0)_{ij}$, $B_{ij}$ and $B'_{ij}$ are listed in Table 7, and the creep-fatigue crack growth properties in air and in a vacuum of the two tested steels are summarized in Fig. 18. In Fig. 19 the fatigue crack growth data for normalized and tempered $2\frac{1}{4}$Cr-1Mo steel in air and in a vacuum (about $1·3 \times 10^{-3}$ Pa) at $525\,°C$ and a test frequency of $10^{-2}$ Hz by Skelton and Challenger [26] are compared with the results shown in Fig. 18. As the test frequency was comparatively low and the type of

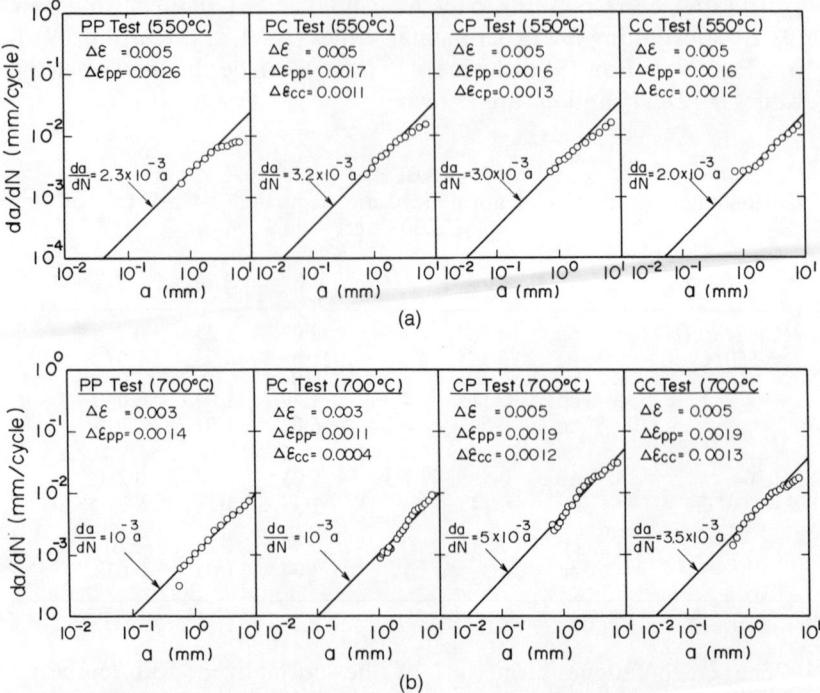

FIG. 17. Crack growth rate versus crack length relationship obtained from *IJ* tests: (a) $2\frac{1}{4}$Cr–1Mo steel (normalized and tempered); (b) SUS 304 steel.

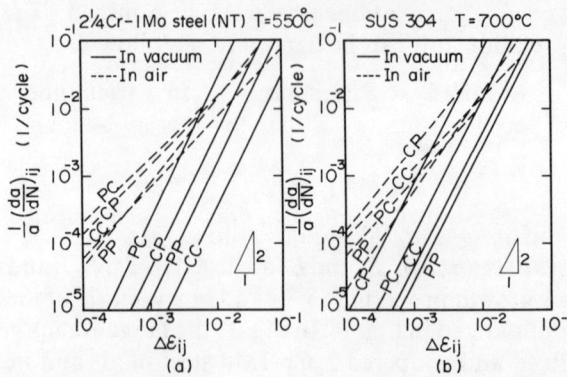

FIG. 18. Summarized illustrations of the creep-fatigue growth properties determined for (a) normalized and tempered $2\frac{1}{4}$Cr–1Mo steel and (b) SUS 304 steels.

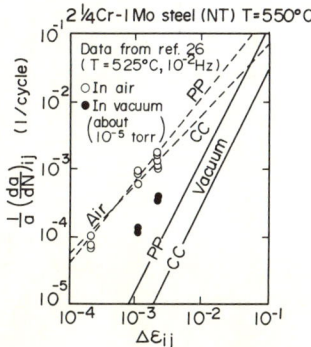

FIG. 19. Comparison between the crack growth properties obtained for normalized and tempered $2\frac{1}{4}$Cr-1Mo steel by the authors and by Skelton and Challenger [26].

cyclic inelastic strain can be regarded as the $\Delta\varepsilon_{pp} + \Delta\varepsilon_{cc}$ type, the data cited were compared with the $(da/dN)_{pp}$ versus $\Delta\varepsilon_{pp}$ relationship and the $(da/dN)_{cc}$ versus $\Delta\varepsilon_{cc}$ relationships shown in Fig. 18.

From the results shown in Fig. 19, it can be said that the crack growth behavior in air can be estimated from crack growth equations based on the proposed model, whereas the fatigue crack growth rate data in a vacuum of $10^{-3}$–$10^{-4}$ Pa were quite different from those estimated from the derived equations in a vacuum. Therefore, it is necessary to conduct the test in a higher vacuum condition in order to confirm the validity of the derived equations for evaluating the creep–fatigue crack growth rate in a perfect vacuum.

It should also be noted in Table 7 that the value of $(a_f/a_0)_{ij}$ tended to be dependent on the kind of material and the type of the inelastic strain range ($\Delta\varepsilon_{ij}$), this fact suggesting that the material-dependent creep-

TABLE 7

Creep-fatigue crack growth properties of normalized and tempered $2\frac{1}{4}$Cr-1Mo steel and SUS 304 stainless steel

| Material | $2\frac{1}{4}$Cr-1Mo (NT) T = 550°C | | | | SUS 304 T = 700°C | | | |
|---|---|---|---|---|---|---|---|---|
| ij | pp | pc | cp | cc | pp | pc | cp | cc |
| $\ln(a_f/a_0)_{ij}$ | 8·3 | 10·3 | 2·6 | 4·0 | 13·7 | 5·5 | 5·4 | 5·2 |
| $B_{ij}$ | 14·0 | 208 | 37·9 | 2·91 | 28·4 | 134 | 843 | 40·6 |
| $B'_{ij}$ | 3·48 | 1·07 | 1·74 | 0·651 | 13·1 | 2·10 | 14·8 | 1·67 |

fatigue fracture behavior could be attributed to the difference in the values of $(a_f/a_0)_{ij}$, or $(a_0)_{ij}$ and $(a_f)_{ij}$, and that further study will be needed on the nucleation mechanism of very small creep–fatigue cracks.

## SUMMARY

The results of the recent study conducted were summarized for the effect of environmental air on the creep–fatigue fracture behavior of austenitic stainless steels and $2\frac{1}{4}$Cr–1Mo steel, based on the SRP concept.

An accelerated testing procedure was presented to obtain the precise SRP life relationships and, using this procedure, the effect of environmental air on the SRP life relationships was experimentally determined for both kinds of materials. It was shown that a vacuum condition such as $10^{-3}$–$10^{-4}$ Pa was not ideal for the SRP life relationships in a vacuum. To obtain the SRP life relationships in a vacuum, the environmental-effect model was presented and the DN-SRP life relationships in a perfect vacuum were derived, which made it possible to evaluate quantitatively the effect of environmental air.

The SRP analysis of creep–fatigue cracks was also presented from the assumption that the SRP life relationships could be closely related to the crack growth properties. The results suggest that further studies along these lines will clarify the nucleation mechanism for very small cracks.

## REFERENCES

[1]  S. S. Manson, G. R. Halford and M. H. Hirschberg, *NASA TMX-67838* (1971).
[2]  S. S. Manson, *ASTM STP 520*, 744 (1973).
[3]  G. R. Halford, M. H. Hirschberg and S. S. Manson, *ASTM STP 520*, 658 (1973).
[4]  S. S. Manson, G. R. Halford and A. C. Nachtigall, *NASA TMX-71737* (1975).
[5]  M. H. Hirschberg and G. R. Halford, *NASA TND-8072* (1976).
[6]  S. S. Manson and G. R. Halford, *1976 ASME-MPC Symp. on Creep–Fatigue Interaction*, ASME, New York, 283 (1976).
[7]  G. R. Halford and S. S. Manson, *ASTM STP 612*, 239 (1975).
[8]  G. R. Halford, J. F. Saltsman and M. H. Hirschberg, *NASA TM-73737* (1977).

[9] K. Hirakawa, K. Tokimasa and K. Toyama, *J. Soc. Mater. Sci. Jpn.*, **27**, 948 (1978).
[10] K. Hirakawa, K. Tokimasa and K. Toyama, *J. Iron Steel Inst. Jpn.*, **65**, 906 (1979).
[11] K. Hirakawa, K. Tokimasa and K. Toyama, *J. Iron Steel Inst. Jpn.*, **65**, 916 (1979).
[12] K. Hirakawa and K. Tokimasa, *J. Soc. Mater. Sci. Jpn.*, **28**, 386 (1979).
[13] K. Hirakawa and K. Tokimasa, *J. Soc. Mater. Sci. Jpn.*, **30**, 65 (1981).
[14] K. Hirakawa and K. Tokimasa, *Sumitomo Search*, **26**, 118 (1981).
[15] K. Tokimasa and I. Nitta, *Sumitomo Search*, **28**, 87 (1983).
[16] K. Tokimasa and I. Nitta, *J. Soc. Mater. Sci. Jpn.*, **35**, 267 (1986).
[17] K. Tokimasa and I. Nitta, *J. Soc. Mater. Sci. Jpn.*, **35**, 274 (1986).
[18] K. Tokimasa, K. Tanaka and I. Nitta, *J. Soc. Mater. Sci. Jpn.*, **35**, 1030 (1986).
[19] J. F. DeLong, R. Ishimoto, K. Tokimasa, A. Ohtomo and T. Honda, Fatigue and Creep Characteristics of Materials for Transportation and Power Industries, *ASME-MPC*, **25**, 173 (1984).
[20] Y. Morita and K. Tokimasa, *J. Iron Steel Inst. Jpn.*, **72**, 210 (1986).
[21] H. D. Solomon and L. F. Coffin, Jr., *ASTM STP 520*, 112 (1973).
[22] L. F. Coffin, Jr., *ASTM STP 520*, 5 (1973).
[23] L. F. Coffin, Jr., *J. Mater.*, **6**, 388 (1971).
[24] L. F. Coffin, Jr., *Metall. Trans.*, **3**, 177 (1972).
[25] R. P. Skelton, *ASTM STP 770*, 337 (1982).
[26] R. P. Skelton and K. D. Challenger, *Mater. Sci. Eng.*, **65**, 271 (1984).

# Creep–Fatigue Damage Evaluation and Estimation of the Remaining Life for Two Heat Resisting Steels

MASATERU OHNAMI and MASAO SAKANE

*Department of Mechanical Engineering, Ritsumeikan University, 56-1 Tojiin Kita-machi, Kita-ku, Kyoto 603, Japan*

*and*

SEIICHI NISHINO

*Graduate School of Ritsumeikan University, 56-1 Tojiin Kita-machi, Kita-ku, Kyoto 603, Japan*

## ABSTRACT

The estimation of remaining life in type 304 stainless steel and $2\frac{1}{4}$Cr–1Mo steel under creep–fatigue interacting conditions was studied using the concept of the remaining life diagram and non-destructive parameters expressing the material damage. The prediction method for the remaining life in creep–fatigue under uniaxial stress was also applied to the case of biaxial stress. The particle size and microstrain obtained by an X-ray profile analysis were effective non-destructive parameters for measuring the material damage under creep–fatigue conditions. The equivalent stress and strain parameters, based on crack opening displacement, also provided an effective comparative basis for estimating the remaining life under biaxial stress.

## INTRODUCTION

Large fossil-fuel power plants and chemical plants with high efficiency have been designed for long-term use. The structural materials used in these applications are exposed to severe environments and must maintain long-term integrity. Some fossil-fuel electricity generating plants have reached their design lives, and strong emphasis has been placed on estimating the remaining life of their structural

165

materials so that the life can be prolonged. To address these problems, a panel discussion entitled 'Remaining life assessment of fossil-fuel power plants — current situation and activities' under the chairmanship of M. Ohnami and M. Kitagawa was held at the International Conference on Creep (April 1986, Tokyo) [1]. In 1983, the Science and Technology Agency in Japan began a 5 year project to develop the fundamental technology for remaining life assessment. The national project entitled 'Development of Evaluation Techniques for the Reliability of Structural Materials' is contributing to these studies [2], and this chapter was written as part of the collaborative study.

When estimating remaining life, it is important to recognize that the stress level attained in practical applications is generally very low, so that destructive testing would be impossible because of the time needed, even if a miniature specimen from the plant component in service could be sampled. To solve this problem, Woodford [3] has proposed the creep remaining life diagram, with which it is possible to estimate the remaining life under low stress levels, and he proposed a determination procedure for the diagram by incremental/decremental stress change tests.

This chapter extends the procedure for the creep remaining life diagram to low cycle fatigue and creep–fatigue diagrams, and is intended for developing non-destructive parameters which can accurately express the creep–fatigue damage, using type 304 austenitic stainless steel and $2\frac{1}{4}$Cr–1Mo ferritic low alloy steel. The drawing method for the creep–fatigue remaining life diagrams is proposed from both the creep and fatigue remaining diagrams, and from the relationship between the remaining life and non-destructive parameters, using X-ray profile analysis and ultrasonic spectrum analysis [4–6]. The estimation for creep–fatigue remaining life under uniaxial stress is extended to that under biaxial stress by using new equivalent stress and strain parameters based on the crack opening displacement (COD) previously proposed by the authors [7–9]. This is examined from the biaxial test data under combined push–pull and reversed torsion, for a thin-walled hollow cylindrical specimen of type 304 stainless steel [10, 11].

## EXPERIMENTAL PROCEDURE

One of the test materials used in this study was a solution of heat-treated type 304 austenitic stainless steel (at 1373 K) with a percentage chemical composition by mass of 0·38 Si, 1·13 Mn, 0·008 P, 0·021 S, 8·8 Ni, 18·52 Cr, 0·06 C and the remainder Fe, and with an ASTM No. 3·5 grain size. The other was a normalized and tempered $2\frac{1}{4}$Cr–1Mo ferritic steel (at 1203 K for 60 min and 963 K for 130 min) with a percentage chemical composition by weight of 0·11 C, 0·25 Si, 0·44 Mn, 0·012 P, 0·004 S, 2·11 Cr, 0·99 Mo and the remainder Fe. Each specimen of the two kinds of heat-resisting steels tested in creep had a diameter of 7 mm and a gage length of 26 mm. The specimens for the fatigue and creep–fatigue tests had a diameter of 10 mm within a gage length of 18 mm. Both these specimens had a polished area of 6 × 10 mm² within the gage length in order to measure the X-ray parameters, 100 $\mu$m of the surface being removed by electropolishing.

The hollow cylindrical specimens, with and without a center notch hole of diameter 1 mm, which were used for the biaxial creep–fatigue and pure fatigue tests had an outer diameter of 12 mm, an inner diameter of 9 mm and a gage length of 15 mm. Each specimen surface was polished with #2000 emery paper before the tests.

In order to draw the creep remaining life diagram of type 304 stainless steel, the creep rupture test was carried out at 873 K in air, and the stress change tests from the original $\sigma_0$ = 255 MPa to $\sigma$ = 294 MPa, 274 MPa or 235 MPa were carried out after creeping for 48% or 75% of the rupture time under $\sigma_0$ = 255 MPa. Stress change tests from $\sigma$ = 294 MPa, 274 MPa or 235 MPa to $\sigma_0$ = 255 MPa were also carried out. To determine the low cycle fatigue remaining life diagram for the materials, as well as that for creep, the total strain range change test was carried out at 873 K, in air, under a fully reversed triangular strain wave (a strain ratio $R$ = −1) in push–pull with a strain rate of 5 × $10^{-3}$ s$^{-1}$. The strain range change tests were carried out from the original $\Delta\varepsilon_{to}$ = 1·0% to $\Delta\varepsilon_t$ = 1·5%, 0·7% or 0·5% after straining for 58% or 74% of the failure life under $\Delta\varepsilon_{to}$ = 1·0%. Strain range change tests from $\Delta\varepsilon_t$ = 1·5%, 0·7% or 0·5% to $\Delta\varepsilon_{to}$ = 1·0% were also carried out. The propriety of the estimation method for creep–fatigue remaining life was examined by carrying out the strain range change tests under a push–pull strain wave with tensile strain hold times $t_H$ of 5 min, 30 min and 24 h. The strain range of $\Delta\varepsilon_{to}$ = 1·0% was changed to $\Delta\varepsilon_t$ = 1·5% or 0·7% after straining for 50% or 70% of the failure life.

Similar stress change and total strain range change tests were carried out on the same size specimens of $2\frac{1}{4}$Cr–1Mo steel at 823 K in air. The creep stress change tests from $\sigma_0$ = 216 MPa to $\sigma_0$ = 235 MPa, 196 MPa or 176 MPa were carried out after creeping for 30%, 50% or 70% of the rupture time under $\sigma_0 = 216$ MPa. The strain range change tests with and without tensile strain hold times, from $\Delta\varepsilon_{to} = 1\cdot0\%$ to $\Delta\varepsilon_t = 1\cdot5\%$, $0\cdot7\%$ or $0\cdot5\%$, were carried out after straining for 30%, 50% or 70% of the failure life under $\Delta\varepsilon_{to} = 1\cdot0\%$.

The applicability of the uniaxial remaining life diagram to biaxial conditions was examined by combined push–pull and reversed torsion low cycle fatigue tests with several principal strain ratios $\phi = \varepsilon_3/\varepsilon_1$, ranging from $-0\cdot5$ to $-1$, using unnotched tubular specimens of type 304 stainless steel at 923 K in air. Mises' equivalent total strain range $\Delta\varepsilon_{eq}$ of $0\cdot7\%$, $1\cdot0\%$ or $1\cdot5\%$ was controlled under the strain ratio of $R = -1$ with a strain rate of $10^{-3}\%$ s$^{-1}$.

The test equipment used was a conventional static creep rupture tester and a combined axial and torsion type of electro-hydraulic servo-controlled fatigue testing machine incorporating a microcomputer [12]. An originally developed long-term creep–fatigue testing machine [13] was used for the 24 h hold time tests. A Rigaku Rotorflex RU-200 was used to conduct the X-ray diffraction experiments, and the ultrasonic attenuation was measured from a longitudinal wave with a central frequency of 10 MHz in the spectrum. The X-ray profile was analyzed with a digitizer of $0\cdot1$ mm resolution power, which was coupled to a microcomputer. This resolution power of $0\cdot1$ mm corresponds to $0\cdot0005°$ on the profile chart. We prechecked the influence of the Gonio scanning speed on the shape of the X-ray profile. From the test results using three scanning speeds of $0\cdot25°$ min$^{-1}$, $0\cdot5°$ min$^{-1}$ and $1°$ min$^{-1}$ we concluded that there was no difference in the shape of the profile due to the Gonio scanning speed. As a result, a scanning speed of $1°$ min$^{-1}$ was employed in the interests of efficiency.

## CREEP, FATIGUE AND CREEP–FATIGUE DAMAGE EVALUATIONS, AND ESTIMATION OF THEIR REMAINING LIVES

Figures 1(a) and 1(b) [4] show the remaining life diagram for type 304 stainless steel in static creep rupture and low cycle fatigue failure, respectively, at 873 K in air, drawn in accordance with the stress change tests already mentioned. In these diagrams, the creep damage factor $D_c$ and the fatigue damage factor $D_f$ were calculated from $D_c = 1 - t/t_R$ and $D_f = 1 - N/N_f$, respectively, where $t_R$ is the creep rupture life under $\sigma_0 = 255$ MPa, $t$ is the remaining life, and $N_f$ and $N$ are the low cycle fatigue failure life under $\Delta\varepsilon_{to} = 1\%$ and the remaining life, respectively.

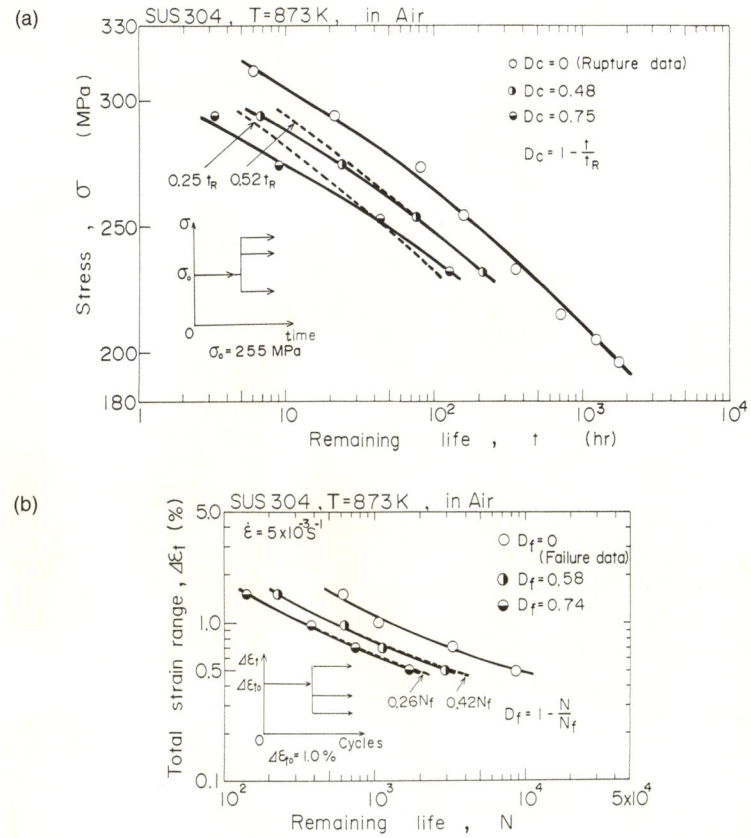

FIG. 1. Remaining life diagrams for type 304 stainless steel under (a) creep and (b) low cycle fatigue at 873 K in air.

$N_f$ is defined as the number of cycles when the tensile stress amplitude decreased to $\frac{3}{4}$ of the maximum value. Therefore, $D_c$ and $D_f$ indicate the creep and fatigue damage to the material in terms of life fractions.

In Fig. 1(a), the broken lines show the damage by the linear damage accumulation rule from the line $D_c = 0$. Since the damage from the test data calculated by the linear damage accumulation rule was in the range 0·89–1·05 (the average damage was 0·91), the linear damage accumulation rule holds approximately for the creep case. However, a significant difference appears between the broken and solid lines at higher stress levels, which shows that a graphical representation like Fig. 1(a) excessively enlarged the difference in remaining life at the

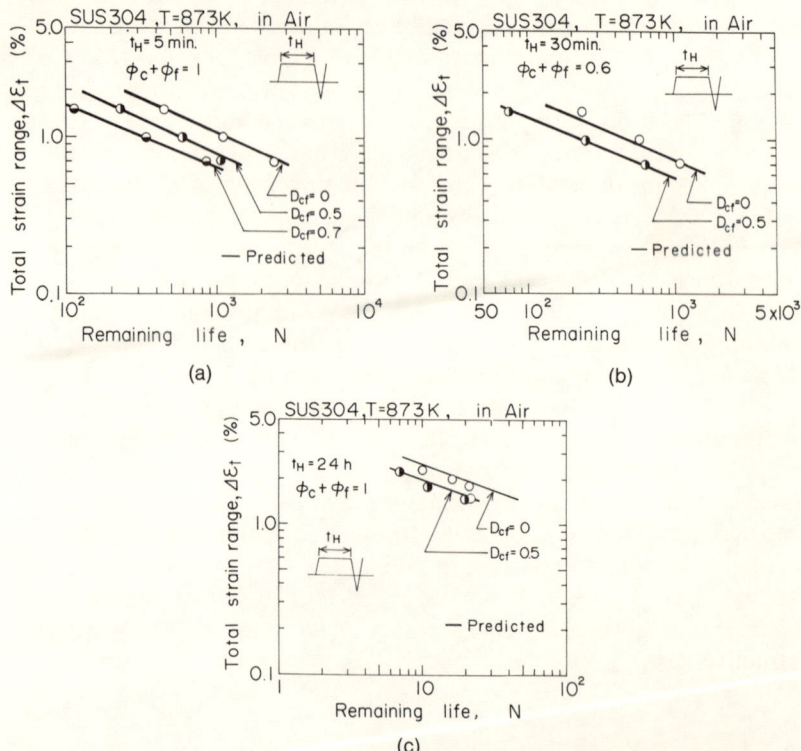

FIG. 2. Remaining life diagrams for type 304 stainless steel under creep–fatigue with (a) $t_H = 5$ min, (b) $t_H = 30$ min and (c) $t_H = 24$ h at 873 K in air.

higher stress level. Figure 1(a) is the result of the stress change test from $\sigma_0 = 255$ MPa to other stress levels, and an almost identical figure was obtained in the stress change test from the other stress levels to 255 MPa, these results not being presented owing to space limitations. Therefore, we could conclude that the remaining life of a material that has been used long term at a low stress level can be obtained from a higher stress test. The remaining life at the high stress level can then be converted to the operating conditions at the lower stress.

In addition to creep, the low cycle fatigue remaining life of the damaged material at $\Delta\varepsilon_{to} = 1\cdot0\%$ can be estimated from Fig. 1(b). The broken lines in the figure were drawn in accordance with the linear damage accumulation rule based on the line $D_f = 0$. The solid lines agree with the broken lines, and the averaged value of the damage from the test data by the linear damage accumulation rule was 0·95, so that the linear damage accumulation rule also holds for the low cycle fatigue case. We obtained similar remaining life diagrams for $2\frac{1}{4}$Cr–1Mo steel at 823 K in air, although the diagrams are not presented here.

Figure 2 [4] shows a comparison between the predicted creep–fatigue remaining lives of type 304 stainless steel under asymmetrical strain waveforms for various hold times; the remaining lives were calculated from both the pure creep and the pure fatigue remaining lives shown in Fig. 1 and the experimental data. In the figures constant damage $D_{cf}$ curves were estimated by the linear damage summation rule, $\phi_c + \phi_f =$ constant, where $\phi_c = \Sigma t/t_r$ and $\phi_f = \Sigma N/N_f$. We found that the predicted remaining lives agreed well with the experimental data for creep–fatigue. We also arrived at the same conclusion for $2\frac{1}{4}$Cr–1Mo steel as shown in Fig. 3 [6]. Therefore, we can state that the creep–fatigue remaining lives of both types of heat-resisting steel can be estimated from the lives in both pure creep and pure fatigue using the linear damage summation rule, if the damage factor $D_{cf}$ of the material under creep–fatigue conditions is known. $D_{cf}$ can be estimated if a miniature specimen from the component is available for the creep–fatigue test, assuming it has a loading history in creep–fatigue. If such a miniature specimen is not available, it is necessary to adopt the subsequent non-destructive tests.

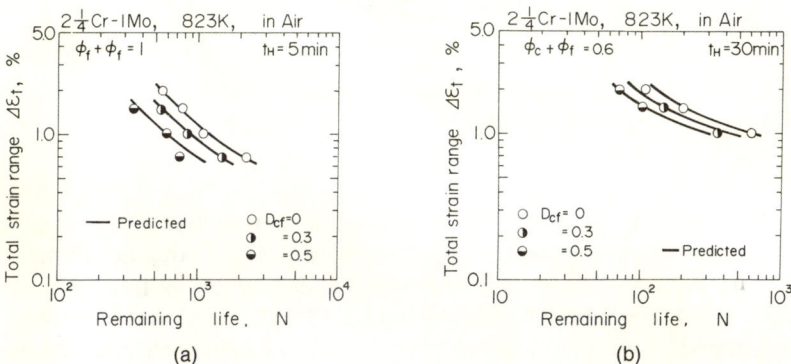

FIG. 3.  Remaining life diagrams for $2\frac{1}{4}$Cr–1Mo steel under creep–fatigue with (a) $t_H$ = 5 min and (b) $t_H$ = 30 min at 823 K in air.

## NON-DESTRUCTIVE EVALUATION OF MATERIAL DAMAGE IN CREEP–FATIGUE

In order to develop a non-destructive evaluation method for material damage in creep–fatigue, it was necessary to find a method that could represent the damage concept, especially the entire process of micro-structural change in the material and deterioration due to distributed microcracks under creep–fatigue conditions. It was also necessary to find a method that could definitively represent the damage concept and measure it with precision in accordance with the elapsed time [14].

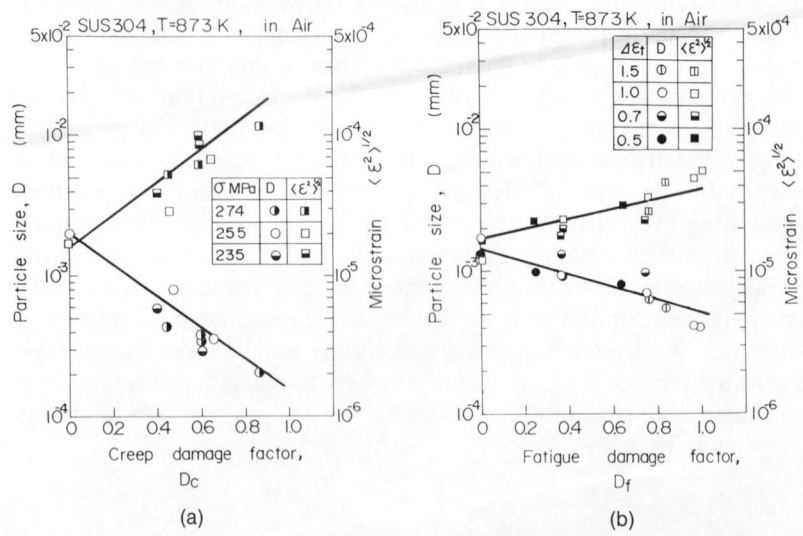

FIG. 4. Experimental relationships of particle size $D$ and microstrain $\langle \varepsilon^2 \rangle^{1/2}$ with (a) creep damage factor $D_c$ and (b) the fatigue damage factor $D_f$ for type 304 stainless steel at 873 K in air.

Figures 4(a) and 4(b) [4] show the experimental relationships between the particle size $D$, the microstrain $\langle \varepsilon^2 \rangle^{1/2}$ and the damage factors $D_c$ and $D_f$ in creep rupture and low cycle fatigue tests, respectively, for type 304 stainless steel at 873 K in air. $D$ and $\langle \varepsilon^2 \rangle^{1/2}$ were calculated from a Fourier analysis of the X-ray profile using the Garrod–Auld method [15]. The particle size $D$ decreased with the increasing creep damage factor $D_c$ while the microstrain $\langle \varepsilon^2 \rangle^{1/2}$ increased with increasing $D_c$ as well as the half-value breadth in the X-ray profile [4]. As particle size is a parameter expressing the cell size or subgrain size in a grain, a dense dislocation structure progressively occurs with creep–fatigue [16]. The microstrain is a parameter to express the lattice microstrain due to dislocation, and increase with increasing $D_c$, which means that the dislocation structure also reflects the microstrain. Also, the particle size linearly decreased with increasing $D_f$, and the microstrain linearly increased with increasing $D_f$, even in the region where the cyclic strain hardening was saturated, in which the half-value breadth in the X-ray profile had an almost constant value and was not appropriate for evaluating $D_f$ [4, 5]. Therefore, we concluded that the particle size and microstrain obtained by X-ray profile analysis were good parameters for evaluating both the pure creep and the pure fatigue damage to the material.

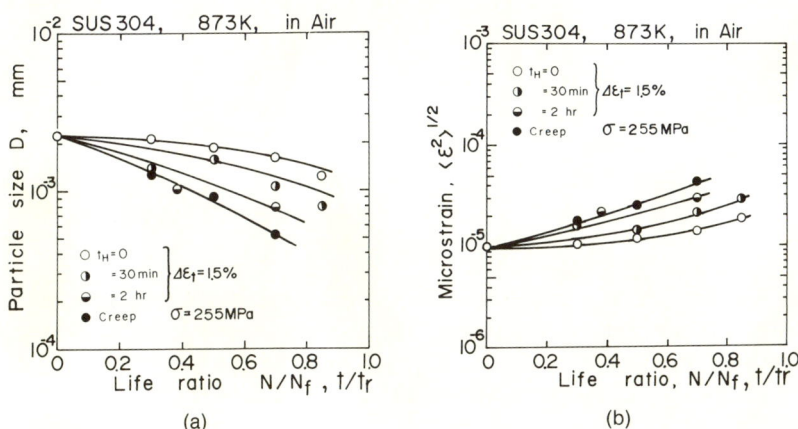

FIG. 5. Experimental relationships of (a) particle size $D$ and (b) microstrain $\langle \varepsilon^2 \rangle^{1/2}$ with the life ratios $N/N_f$ and $t/t_r$ for type 304 stainless steel at 873 K in air.

Figure 5(a) and 5(b) [4] are plots of $D$ and $\langle \varepsilon^2 \rangle^{1/2}$ as a function of the life ratios $N/n_f$ and $t/t_r$, in the constant strain range test with and without hold times and in the constant stress creep test, respectively, for type 304 stainless steel at 873 K in air. The data for both $t_H = 30$ min and $t_H = 2$ h express the intermediate values between the pure creep and pure fatigue data. In all the cases indicating the creep–fatigue condition, since a clear correlation held between the X-ray parameters and life ratios, the X-ray parameters were effectively non-destructively measuring the damage to the material. Figure 6(a) and 6(b) [6] also show a similar relationship between $D$, $\langle \varepsilon^2 \rangle^{1/2}$ and the life ratios $N/N_f$ and $t/t_r$, respectively, for $2\frac{1}{4}$Cr-1Mo steel at 823 K in air. Thus, we can draw the important conclusion that estimating the creep–fatigue remaining life of structural materials is possible from both the remaining life diagrams using the non-destructive X-ray parameters when the loading history of the materials is known.

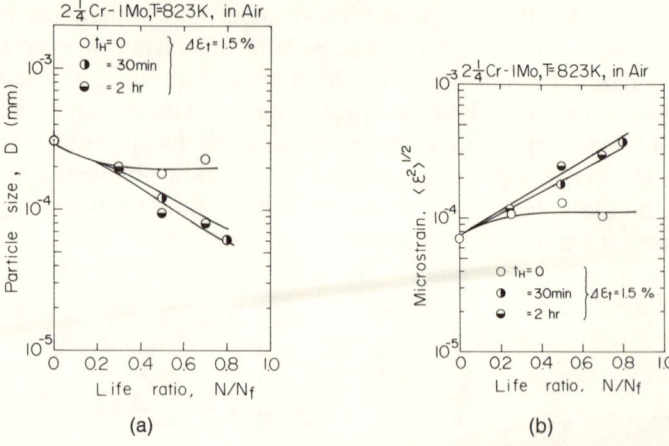

FIG. 6. Experimental relationships of (a) particle size $D$ and (b) microstrain $\langle\varepsilon^2\rangle^{1/2}$ with the life ratios $N/N_f$ and $t/t_r$ for $2\frac{1}{4}$Cr-1Mo steel at 873 K in air.

We shall briefly describe here the correlation of the X-ray parameters with the parameters obtained from another non-destructive testing method. Figures 7(a), 7(b) and 7(c) [4, 6] show the experimental relationships of the ultrasonic attenuation coefficient $\alpha$ with the density change $-\Delta\rho/\rho$, the particle size $D$ and the microstrain $\langle\varepsilon^2\rangle^{1/2}$, respectively, in the total strain-controlled push–pull tests with and without hold time, and in the pure creep test, for the two kinds of steel tested. The data for type 304 stainless steel are plotted as circles and those for $2\frac{1}{4}$Cr–1Mo steel as squares. $\alpha$ was calculated from the equation $\alpha = (20/2l)\log(A_1/A_2)$, where $A_1$ and $A_2$ are the areas of the spectra (amplitude versus frequency curve) from the first and second reflections, respectively, in the ultrasonic test, and $l$ is the length of the specimen injected. Since it is generally more difficult to transmit an elastic wave in an austenitic steel than in a ferritic steel, the scatter in the data for the height of the ultrasonic reflection wave is larger for austenitic steel. Therefore, we used here the ratio of the energy value, $A_1/A_2$, instead of that of the reflected wave height. The suitability of ultrasonic noise energy values for non-destructive damage evaluation has been proved by other researchers [17] using a power boiler superheater tube of type 316 stainless steel (SUS 316 HTB) and $2\frac{1}{4}$Cr–1Mo steel (STBA24) under creep or aging conditions for over 50·000 h. Both end surfaces of the specimen were polished with emery paper because the echo pattern was influenced by the roughness of the end surface. Measurement of the

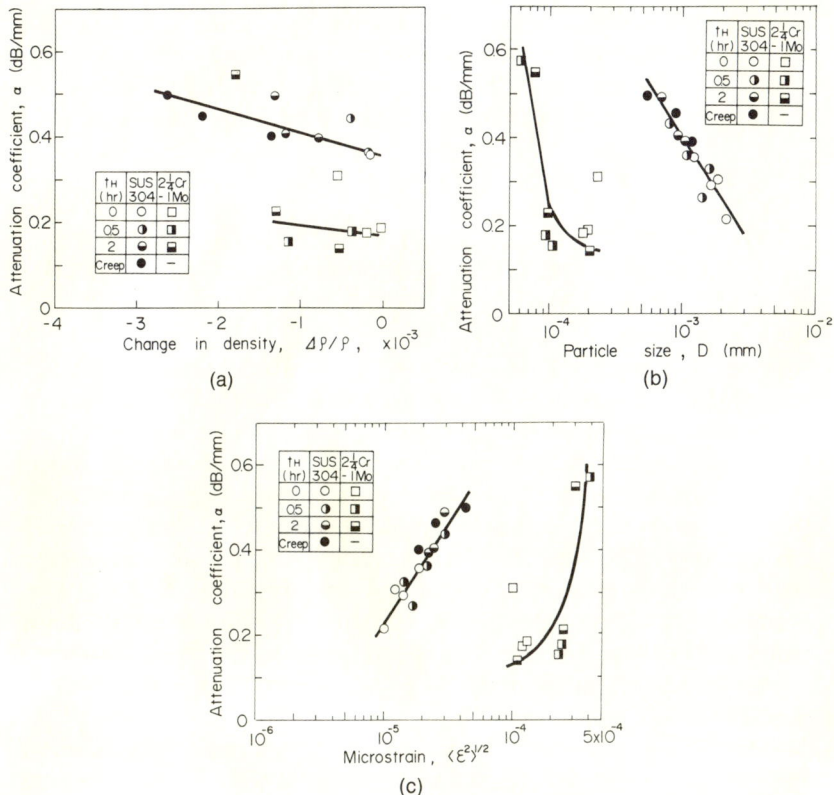

FIG. 7. Experimental correlations between the ultrasonic attenuation coefficient $\alpha$ and (a) the density change $-\Delta\rho/\rho$, (b) the particle size $D$ and (c) the microstrain $\langle\varepsilon^2\rangle^{1/2}$ for both type 304 stainless steel and $2\frac{1}{4}$Cr-1Mo steel.

density change was made by a specific gravity balance method (JIS Z-8807) with a minimum scale of $10^{-4}$ g.

We found from the figures that the increase of the ultrasonic attenuation coefficient in terms of the energy value did not correlate well with the density changes, but did correlate well with both the decrease of particle size and the increase of microstrain, irrespective of the load waves tested. This suggests that the change of ultrasonic attenuation in terms of the energy value tested will be mainly a result of the change in dislocation structure and dislocation density of the metals under creep–fatigue conditions [16]. Therefore, we concluded that the X-ray parameters of the particle size and the microstrain, as well as the ultrasonic attenuation coefficient in terms of the energy value, were effective measures of material damage in creep, fatigue and creep-fatigue, irrespective of the transgranular and intergranular fracture modes. This has been confirmed in a previous paper by the authors [18] from X-ray fractography in creep–fatigue of type 304 stainless steel.

## APPLICATION OF THE ESTIMATION METHOD FOR CREEP-FATIGUE REMAINING LIFE UNDER UNIAXIAL STRESS TO THE BIAXIAL STRESS CASE

In previous papers by the authors [10], we found that, for a strain base, the $\Gamma$ plane parameter [19], $\Gamma^*$ plane parameter [20], COD strain $\varepsilon^*$ [7-10] and maximum principal strain $\varepsilon_1$ were effective for correlating the high temperature, low cycle biaxial fatigue failure life data. However, Mises' strain $\varepsilon_{eq}$ and maximum shear strain $\gamma_{max}$ yielded a poor correlation. As to the stress parameter, only the COD stress $\sigma^*$ gave a satisfactory result, and Mises' stress $\sigma_{eq}$, the principal stress $\sigma_1$ and the maximum shear stress $\tau_{max}$, yielded a poor correlation. This was also examined for the problem of creep–fatigue [11] and for that of strain ratios $\phi = \varepsilon_3/\varepsilon_1$ ranging from $-1$ to $1$ [21]. Therefore, in this study we applied the COD stress/strain range to estimating the creep–fatigue remaining life under biaxial stress.

Figure 8(a) and 8(b) [10] show the respective correlations of the low cycle biaxial fatigue failure data for the unnotched hollow cylindrical specimen at 923 K in air with the COD strain and stress ranges calculated from Eqns. (1) and (2) [7, 10]:

$$\varepsilon_1^* = C\varepsilon_1(\phi^2 + \phi + 1)^{1/2}[(\phi + 2)\{3(\phi^2 + \phi + 1)\}^{-1/2}\{3/(\phi + 2)\}^m]^{1/n} \quad (1a)$$

where $C = 2^{(n-m)/n}/\sqrt{3}$, and

$$\varepsilon_{II}^* = 0.5|\gamma_{max}| \quad (1b)$$

where $\varepsilon_1^*$ and $\varepsilon_{II}^*$ are the COD strains for crack mode I (macrocrack propagation perpendicular to the maximum principal stress) and crack mode II (in the direction of the maximum shear stress), respectively. In this equation, $m$ is a material constant which takes the value $0.5$ for $\varepsilon_3/\varepsilon_1$ values ranging from $-1$ to $-\nu$, for type 304 stainless steel, $n$ is the strain hardening exponent of the material, and $\nu$ is Poisson's ratio which takes the value $0.5$ in creep–fatigue. The constants $n$ and $C$ have the respective values of $0.56$ and $0.62$ for a type 304 stainless steel.

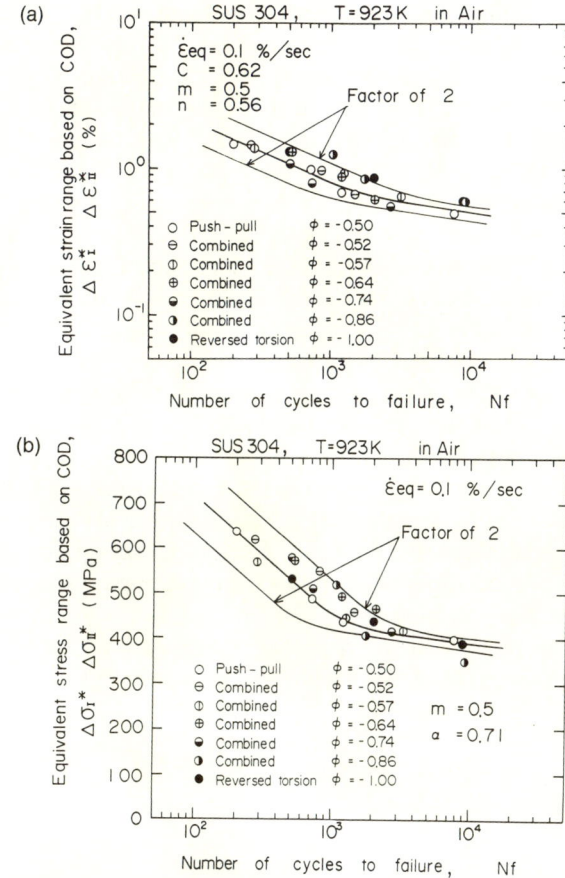

FIG. 8. Arrangement of low cycle biaxial fatigue failure data for type 304 stainless steel (unnotched specimen) at 923 K in air with (a) the COD strain range $\Delta\varepsilon^*$ and (b) the COD stress range $\Delta\sigma^*$.

Equation (1) was originally derived from the following equation for the stress base, by assuming the relationship $\varepsilon^* = \sigma^{*1/n}$ [7, 10]:

$$\sigma_I^* = \alpha\sigma_1(2 - \lambda)^m \qquad (2a)$$

where $\lambda = \sigma_3/\sigma_1$ and $\alpha = 1/2^m$, and

$$\sigma_{II}^* = 1\cdot4\,|\tau_{max}| \qquad (2b)$$

where $\sigma_I^*$ and $\sigma_{II}^*$ are COD stresses for crack modes I and II, respectively.

As it was found that the crack transition from mode I to mode II occurred between $\varepsilon_3/\varepsilon_1 = -0.6$ and $\varepsilon_3/\varepsilon_1 = -0.7$ for the material tested [10, 22], we adopted either the mode I or the mode II parameter according to the value of $\varepsilon_3/\varepsilon_1$. We found from the figures that the low cycle biaxial fatigue failure life correlated well with the push–pull one using the COD stress/strain parameter, in which the effect of the principal stress parallel to the macrocrack on the crack propagation rate was taken into consideration [7, 23].

Figure 9 [11] shows the failure life data correlation of push–pull tests having tensile strain hold times $t_H$ of 0, 10, 30 and 60 min with reversed torsion tests having the same hold times using Eqn. (2). The tests were carried out on the notched hollow cylindrical specimen of type 304 stainless steel at 923 K in air. We found from the figure that the COD stress range $\sigma^*$ could be successfully applied to the low cycle fatigue data even in the hold time tests. This means that Fig. 9 also exhibits the creep–fatigue remaining life diagram for $D_{cf} = 0$ under a biaxial stress state, if the abscissa $N_f$ is replaced by the remaining life. Therefore, we concluded that the remaining creep–fatigue life diagram shown in Fig. 2 could be successfully applied to the case of biaxial stress by using COD stress/strain parameters.

Note that the COD stress in Eqn. (2) was derived from a Finite Element Method analysis, so that the result may be influenced by the

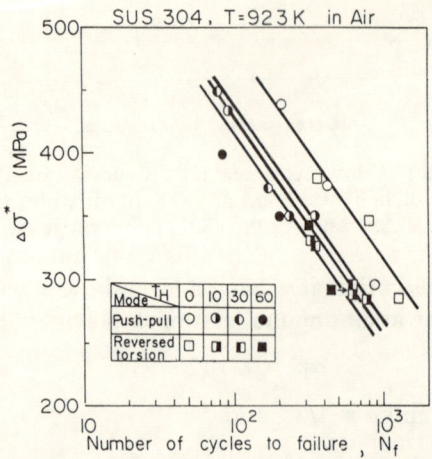

FIG. 9. Arrangement of push–pull and reversed torsion test data in creep–fatigue for type 304 stainless steel (notched specimen) at 923 K in air with the COD stress range $\Delta\sigma^*$.

material constants, i.e. the constitutive equation, $\varepsilon_{eq} = (\sigma_{eq}/E') + B(\sigma_{eq}/\sigma_y)^{n'}$, where $\sigma_{eq}$ and $\varepsilon_{eq}$ are an equivalent stress and strain of the Mises type, $E'$ is the tangent modulus in the elastic part of the $\sigma_{eq}$ versus $\varepsilon_{eq}$ curve, $n'$ is the strain hardening exponent, and $\sigma_y$ is the yield stress. In this regard, the authors have confirmed that the value of $B$ in the constitutive equation does not affect the value of $m$, but that the value of $n$ does [8]. For $1Cr-1Mo-\frac{1}{4}V$ rotor steel, whose $n$ value of 0·2 is smaller than that of a type 304 stainless steel, the value of $m$ is 1·0 when $-1 \leqslant \sigma_3/\sigma_1 < 0$ and is 0·15 when $0 < \sigma_3/\sigma_1 \leqslant 1$ at 823K in air [21]. The appropriate $m$ value should be used for the respective material.

## CONCLUSIONS

(1) Creep and low cycle fatigue remaining life diagrams were produced for a type 304 stainless steel and a $2\frac{1}{4}Cr-1Mo$ steel at 873 K and 823 K, respectively, in air. If a miniature specimen is available from the structural component for mechanical tests, the remaining life can be obtained by converting the test results at high stress levels into those at a lower stress using the remaining life diagram.

(2) Estimation of the creep–fatigue remaining lives of the two steels for tensile strain hold times of 5 min, 30 min and 24 h was accomplished by using both the remaining life diagrams for pure creep and pure fatigue and the linear damage summation rule.

(3) Under creep–fatigue conditions, the microstrain and particle size obtained from an X-ray profile analysis, as well as the ultrasonic attenuation coefficient in terms of the energy value, were applicable non-destructive parameters for estimating the creep–fatigue damage factor, if the loading stress or strain wave form was known. This is effective for estimating non-destructively the creep–fatigue remaining life of the materials.

(4) Creep–fatigue remaining life diagrams under uniaxial stress can be successfully applied to the biaxial stress state using the COD stress/strain parameter, in which the effect of the principal stress parallel to the macrocrack on the creep–fatigue crack propagation rate is considered.

## ACKNOWLEDGMENTS

The authors wish to thank M. Sawada, a graduate student of Ritsumeikan University, for performing part of the experiments. This work was supported by the Science and Technology Agency of Japan as part of the national project entitled 'Development of Evaluation Techniques for the Reliability of Structural Materials'.

## REFERENCES

[1] Panel discussion 'Remaining life assessment of fossil fuel power plants — current situation and activities', *Int. Conf. on Creep, Tokyo, April 16, 1986,* The Japan Society of Mechanical Engineers, Tokyo (1986).

[2] *Report of the Study on Development of Evaluation Techniques for the Reliability of Structural Materials in 1983–1985,* The Science and Technology Agency, Tokyo (1987).

[3] D. A. Woodford, *J. Eng. Mater. Technol., Trans. ASME,* **101**, 311 (1979).

[4] S. Nishino, M. Sakane and M. Ohnami, *Proc. 30th Japan Congr. on Materials Research,* The Society of Materials Science, Japan, 61 (1987).

[5] M. Ohnami, M. Sakane and S. Nishino, *Proc. Int. Conf. on Role of Fracture Mechanics in Modern Technology,* G. C. Sih and H. Nishitani, Eds., North-Holland, Amsterdam, 369 (1987).

[6] S. Nishino, M. Sakane and M. Ohnami, *Proc. 24th Symp. on Strength of Materials at High Temperatures,* The Society of Materials Science, Japan, 87 (1987).

[7] N. Hamada, M. Sakane and M. Ohnami, *Bull. Jpn. Soc. Mech. Eng.,* **28**, 1341 (1985).

[8] N. Hamada, M. Sakane and M. Ohnami, *Proc. 30th Jpn. Congr. on Materials Research*, The Society of Materials Science, Japan, 69 (1987).

[9] N. Hamada, M. Sakane and M. Ohnami, *J. Eng. Mater. Technol., Trans. ASME,* **109** (1987).

[10] M. Sakane, M. Ohnami and M. Sawada, *J. Eng. Mater. Technol., Trans. ASME,* **109**, 236 (1987).

[11] N. Hamada, M. Sakane and M. Ohnami, *Fatigue Eng. Mater. Struct.,* **7**, 85 (1985).

[12] M. Ohnami and N. Hamada, *Proc. 24th Jpn. Congr. on Materials Research,* The Society of Materials Science, Japan, 93 (1982).

[13] M. Sakane, M. Ohnami and Y. Awaya, *Preprint of 34th Annu. Meet. of the Society of Materials Science, Japan,* 73 (1985).

[14] M. Ohnami, Panel discussion 'Remaining life assessment of fossil fuel power plants — current situation and activities, *Int. Conf. on Creep, Tokyo, April 16, 1986,* The Japan Society of Mechanical Engineers, Tokyo, 1 (1986).

[15] R. I. Garrod and J. H. Auld, *Acta Metall.,* **3**, 190 (1955).

[16] S. Nishino, S. M. Sakane and M. Ohnami, *Proc. 29th Jpn. Congr. on Materials Research*, The Society of Materials Science, Japan, 53 (1986).

[17] M. Nakashiro, H. Yoneyama and A. Ohtomo, *Report of the Study on Development of Evaluation Techniques for the Reliability of Structural Materials in 1983–1985,* The Science and Technology Agency, Japan, 30 (1987).

[18]  S. Nishino, M. Sakane and M. Ohnami, *Proc. 29th Jpn. Congr. on Materials Research*, The Society of Materials Science, Japan, 61 (1986).
[19]  M. W. Brown and K. J. Miller, *Fatigue Eng. Mater. Struct.*, 1, 217 (1979).
[20]  R. D. Lohr and E. G. Ellison, *Fatigue Eng. Mater. Struct.*, 3, 1 (1980).
[21]  M. Ohnami, M. Sakane, S. Nishino and T. Itsumura, *Preprint of Japan-U.S. Joint Seminar on Materials Needs for Severe Service Conditions, May 19–23, Tokyo, 1986*, The Japan-U.S. Cooperative Science Program, 149; *Advanced Materials for Severe Service Applications*, K. Iida and A. J. McEviley, Eds., Elsevier Applied Science, Barking, in press.
[22]  A. Nitta, T. Ogata and K. Kuwabara, *Proc. 23rd Symp. on Strength of Materials at High Temperatures*, The Society of Materials Science, Japan, 127 (1985).
[23]  M. Ohnami and M. Sakane, *Inter. J. Eng. Fracture Mech.*, 19 (1987).

# Low Cycle Fatigue Strength and its Prediction for Dissimilar-Metal Electron-Beam-Welded Joints at High Temperature

MASAKAZU OKAZAKI and YOSHIHARU MUTOH

*Department of Mechanical Engineering, Technological University of Nagaoka, Tomioka, Nagaoka, Japan*

and

YOSHIYASU ITOH

*Heavy Apparatus Engineering Laboratory, Toshiba Corporation, Turumi-ku, Yokohama, Japan*

*ABSTRACT*

Low cycle fatigue tests of dissimilar-metal electron-beam-welded joints (DMEBWJs) between A387 Gr.22 steel and SUS 405 steel, and between A387 Gr.22 steel and SUS 316 steel, were carried out with strain-controlled cycling over the welded joints at a temperature of 839 K. It was found that the fatigue lives of the DMEBWJs were significantly shorter than those of the base metals. This resulted from the strain concentration on the material with the lower deformation resistance. It was also qualitatively found that there was an inverse proportionality relationship between the strain distribution and the hardness distribution. Furthermore, a simple prediction method for the low cycle fatigue strength of DMEBWJs was proposed by representing the cyclic hardening and softening properties of the base metals in the joint. The calculated variation of strain distribution on the base metal side of the DMEBWJ with strain cycling agreed with the experimental results. The predicted fatigue lives were also in good agreement with the experimental results, except for the case in which the strain concentration occurred on a heat-affected zone with lower hardness than the base metals.

## INTRODUCTION

Dissimilar-metal welded joints (DMWJs) in service at elevated temperatures are subjected to thermal stress cycling resulting from start-

183

up, shut-down and load change [1, 2]. To enable the fatigue life of these welded joint to be predicted, many studies on low cycle fatigue strength at elevated temperatures have been performed [1–9]. However, the fatigue fracture behaviour of DMWJs is very complicated, because of the discontinuity of mechanical properties and of metallurgical inhomogeneities. Consequently, an adequate design standard has not yet been established. For example, the *ASME Boiler and Pressure Vessel Code, Section III,* has only recommended that the accumulated strain of the weldment must be less than half that of the base metal.

In this work, the low cycle fatigue strength of dissimilar-metal electron-beam-welded joints (DMEBWJs) was investigated. Using a specimen on whose surface a heat-resisting moiré grid [10] was inscribed, the local strain distribution in DMEBWJs could be measured, and the relationship between this strain distribution and the fatigue fracture behaviour is discussed. Furthermore, a simple prediction method for the fatigue life of DMEBWJs is proposed by representing the cyclic hardening and softening properties of the base metals in the joint.

## EXPERIMENTAL PROCEDURE

The base metals used were SCMV4 steel (ASTM A387, Gr.22), SUS 405 ferritic stainless steel (AISI 405) and SUS 316 austenitic stainless steel (AISI 316). In this work, two kind of SUS 405 ferritic stainless steels whose chemical compositions and forging conditions were a little different were used. These base metals are denoted by A387, SUS 405, SUS 405′ and SUS 316, respectively. Their chemical compositions and conditions of heat treatments are given in Table 1, these heat treatments being in accordance with the Japan Industrial Standard [11]. Three kinds of DMEBWJs were examined in this work: a joint between A387 and SUS 405 (A387–SUS 405), one between A387 and SUS 405′ (A387–SUS405′) and one between A387 and SUS 316 (A387–SUS 316). The electron beam welding conditions are given in Table 2, and the width of weld metal in each joint was about 2 mm. A specimen whose geometry is shown in Fig. 1 was taken from the weldment, so that the weld metal was central in the specimen gage length. The specimen shown in Fig. 1(a) was used to evaluate the fatigue strength of each DMEBWJ.

Low cycle isothermal fatigue tests were carried out with the strain-controlled cycling over the welded joint by means of servo-electro-hydraulic test equipment at a temperature of 839 K in air. The test

TABLE 1
Chemical composition (wt. %) and heat treatment of the base metals

| Material | C | Si | Mn | P | S | Ni | Cr | Mo | Al |
|---|---|---|---|---|---|---|---|---|---|
| A387 | 0·14 | 0·03 | 0·54 | 0·006 | 0·006 | — | 2·38 | 1·04 | — |
| SUS 316 | 0·06 | 0·49 | 0·83 | 0·028 | 0·003 | 11·67 | 16·59 | 2·16 | — |
| SUS 405 | 0·06 | 0·34 | 0·56 | 0·027 | 0·003 | 0·15 | 13·19 | — | 0·19 |
| SUS 405' | 0·07 | 0·61 | 0·42 | 0·027 | 0·014 | — | 11·98 | — | — |

*Heat treatment*
A387: 1203 K × 80 min, air cooled; 963 K × 90 min, air cooled.
SUS 316: 1373 K × 10 min, water cooled.
SUS 405: 1073 K × 3 h, air cooled.
SUS 405': 1073 K × 2 h, furnace cooled.

conditions are summarized in Table 3. The specimen was heated by means of a high frequency induction heating system. The temperature difference along the specimen gage length was less than 5 K. The strain wave form used was a fully reversed symmetrical triangular one of the push–pull type. The axial strain was controlled by a strain detector attached to the specimen surface, which contained a linear voltage differential transformer. The repeated strain rate was high enough that the creep effect was negligible [12]. The fatigue life $N_f$ is defined as the

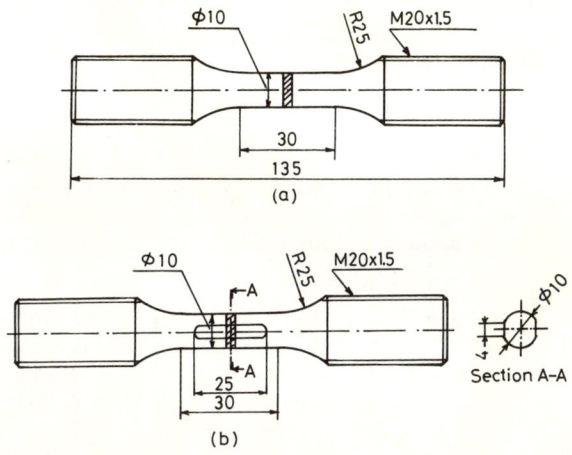

FIG. 1. Geometry of specimens used (dimensions in millimetres): (a) specimen for life evaluation; (b) specimen for measuring strain distribution.

TABLE 2

Conditions for electron beam welding and post-weld heat treatment

| Dissimilar-metal welded joint | Welding voltage (kV) | Welding current (mA) | Welding speed (mm min⁻¹) | Vacuum (Torr) |
|---|---|---|---|---|
| A387–SUS 405 A387–SUS 405' | 55 | 160 | 400 | $<10^{-4}$ |
| A387–SUS 316 | 50 | 227 | 400 | $<10^{-4}$ |

*Post-weld heat treatment*
A387–SUS 405, A387–SUS 405'; 963 K × 90 min, furnace cooled.
A387–SUS 316; 863 K × 3 h, furnace cooled

number of cycles corresponding to a tensile load reduction of 25% from the stationary value.

In order to measure the local strain distribution in the DMEBWJ, the gauge section of the specimen shown in Fig. 1(a) was machined as shown in Fig. 1(b), and a moiré grid of 500 lines in⁻¹ was inscribed on the plane surface of the specimen, and then the specimen was electro-polished. The procedure of electropolishing converted the moiré grid into a heat-resisting moiré grid [10]. The local strain distribution to the loading direction was measured by the dimensional change of the moiré grid, based on the maximum compressive strain, after the specimen had been unloaded at the maximum strain subsequent to the prescribed strain cycling. This measurement was carried out at strain cyclings of about $0.1N_f$, $0.3N_f$, $0.5N_f$ and $0.7N_f$. The details of the measurement procedure are described elsewhere [6, 7].

TABLE 3

Low-cycle fatigue testing conditions

| Material | Temperature (K) | Strain rate (s⁻¹) | Total strain range |
|---|---|---|---|
| A387 | 839 | $10^{-3}$ | 0·36, 0·69, 0·97, 1·57, 2·32 |
| SUS 316 | | | 0·37, 0·68, 0·95, 1·47, 1·95 |
| SUS 405 | | | 0·50, 0·72, 1·00, 1·54, 2·40 |
| SUS 405' | | | 0·50, 0·99, 1·54 |
| A387–SUS 316 | | | 0·35, 0·50, 0·70, 1·00, 1·50 |
| A387–SUS 405 | | | 0·20, 0·34, 0·67, 0·82, 1·00 |
| A387–SUS 405' | | | 0·30, 0·48, 0·99 |

# LOW CYCLE FATIGUE FRACTURE BEHAVIOUR OF DMEBWJS

## Vickers Hardness Distribution

The hardness distributions of the DMEBWJs are given in Figs. 2–4. The hardness of the weld metals was higher than those of the base metals in all cases. The hardness on the A387 side was also higher than those of the other base metals in all cases. The scatter of hardness on the SUS 405 side in Fig. 2 resulted from the indentation position of the

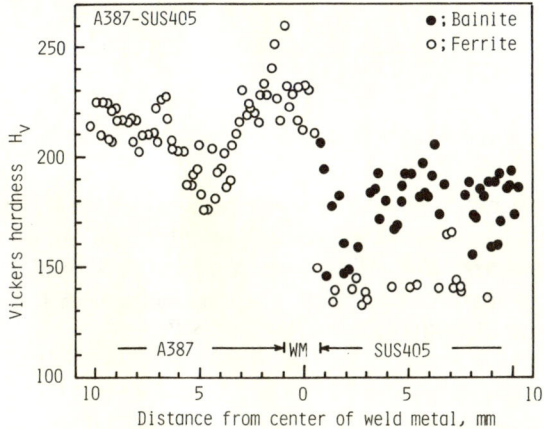

FIG. 2. Hardness distribution for A387–SUS 405.

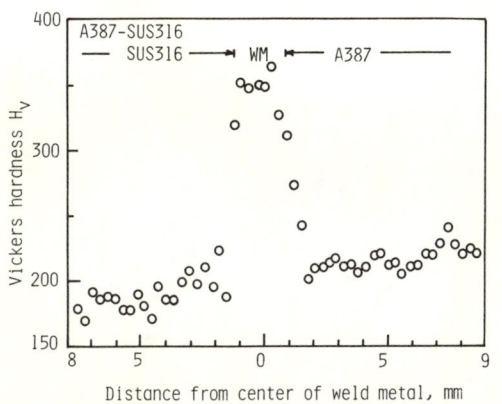

FIG. 3. Hardness distribution for A387–SUS 316.

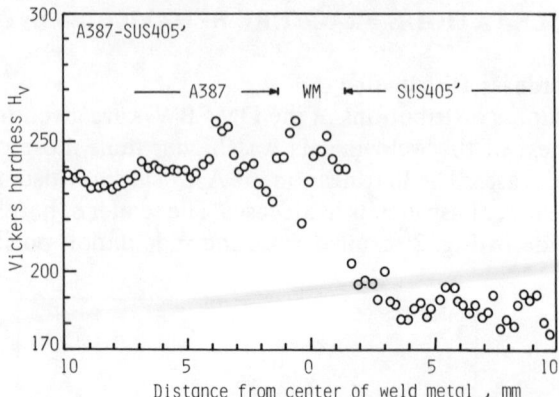

FIG. 4. Hardness distribution for A387–SUS 405'.

micro-Vickers hardness testing machine being in the ferritic phase (○)
or bainitic phase (●), because the metallurgical microstructure of
SUS 405 was coarse. Regarding the hardness of the bainitic phase, it was
found that a heat-affected zone of about 2 mm with lower hardness
existed on the SUS 405 side in A387–SUS405.

**Fatigue Life**

The relationships between the total strain range $\Delta\varepsilon_t$ and the fatigue
life $N_f$ of both the base metals and the DMEBWJs are shown in Fig. 5,
indicating that the fatigue life of each DMEBWJ was significantly
shorter than those of the base metals.

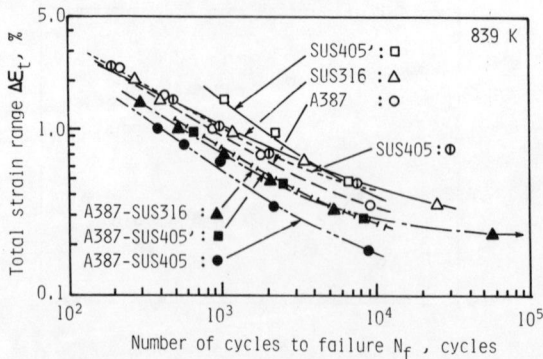

FIG. 5. Fatigue lives of base metals and DMEBWJs.

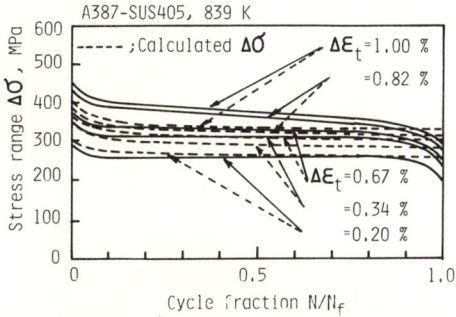

FIG. 6. Variation of stress range with strain cycling in A387-SUS 405.

In Figs. 6–8, the variation of stress range $\Delta\sigma$ with strain cycling in each joint is given. From Fig. 7 it can be seen in A387–SUS 316 that $\Delta\sigma$ increased at an early stage and then decreased with strain cycling. On the other hand, $\Delta\sigma$ in A387–SUS 405 and A387–SUS 405' decreased monotonically with strain cycling (Figs. 6 and 8). As will be subsequently described in detail, SUS 405, SUS 405' and A387 cyclically softened, while SUS 316 cyclically hardened. Such behavior of the base metals is reflected in the $\Delta\sigma$ value of the joints shown in Figs. 6–8. The broken lines in these figures will be described later.

Although the $\Delta\sigma$ values of the joints varied with strain cycling, $\Delta\sigma$ in each DMEBWJ produced a stationary value at $N_f/2$ (where $N_f$ is the fatigue life). The relationships between $\Delta\sigma$ and the plastic strain range at $N_f/2$ are given in Fig. 9. The deformation resistances of the base

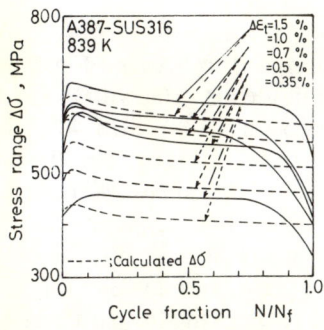

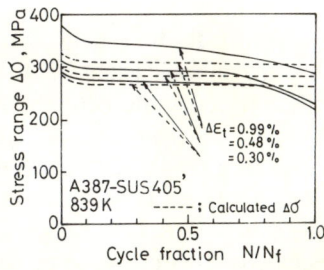

FIG. 7. Variation of stress range with strain cycling in A387–SUS 316.

FIG. 8. Variation of stress range with strain cycling in A387–SUS 405'.

metals increased in the order SUS 405′, SUS 405, A387 and SUS316. From Fig. 9 it is also found that the deformation resistance of the joints was the median of those of the base metals.

The macroscopic photographs of the specimens after testing are given in Fig. 10. The DMEBWJs ruptured on the SUS 405 side in A387–

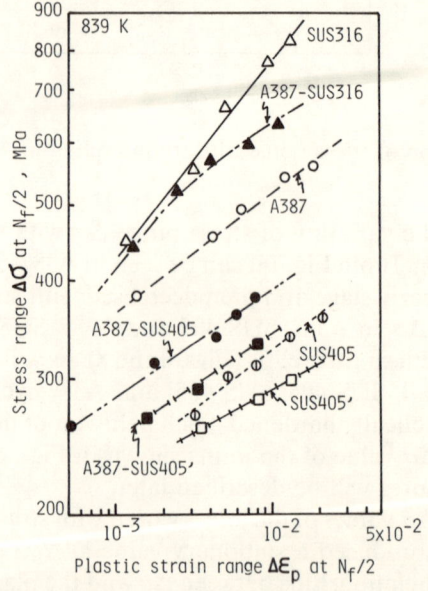

FIG. 9. Cyclic stress–strain relationships for base metals and DMEBWJs.

SUS405, at the bond region of the SUS 405′ side in A387–SUS 405′, and on the A387 side in A387–SUS 316. A comparison between Fig. 9 and Fig. 10 indicates that the joints ruptured in the material with the lower deformation resistance. In A387–SUS 405 and A387–SUS 405′, it is worth noting that these joints ruptured on the SUS 405 or SUS 405′ side, although the fatigue lives of SUS 405 and SUS 405′ are longer than that of A387 (Fig. 5). In A387–SUS 405, the rupture point depended on $\Delta\varepsilon_t$ (where $\Delta\varepsilon_t$ is the total strain range); it was on the base metal side far away from the weld metal in a high strain range, and it was in the heat-affected zone with the lower hardness (Fig. 2) in the low strain range.

Figure 11 shows the variation of average hardness on both sides of the base metals of A387–SUS 316 with strain cycling. The hardness of A387

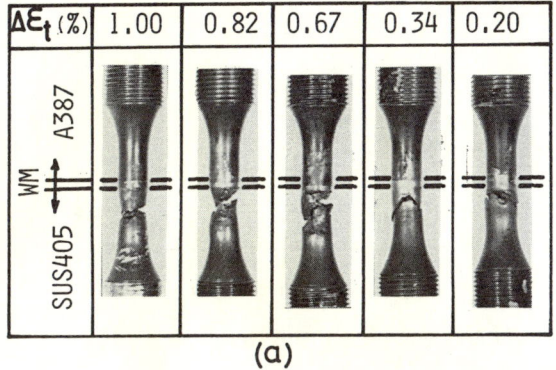

(a)

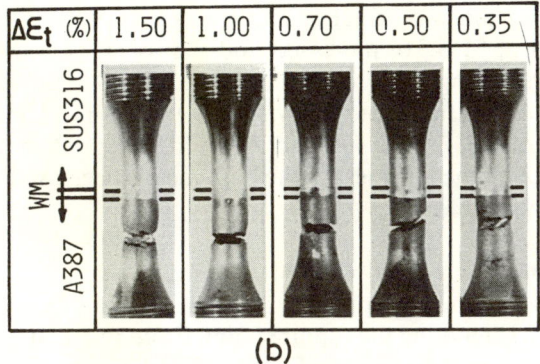

(b)

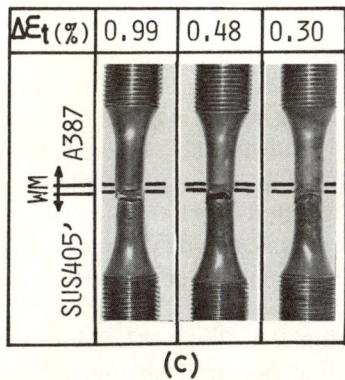

(c)

Fig. 10. Macroscopic rupture photographs of DMEBWJs: (a) A387–SUS 405; (b) A387–SUS 316; (c) A387–SUS 405'.

192          M. OKAZAKI, Y. MUTOH AND Y. ITOH

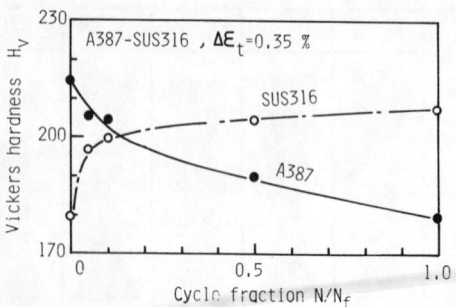

FIG. 11. Variation of hardness with strain cycling in A387–SUS 316.

was higher than that of SUS 316 at an early stage of strain cycling, before the hardness of SUS 316 increased and that of A387 decreased with further strain cycling. As a result the hardness of SUS 316 was higher than that of A387 after about $0.15N_f$. On the other hand, the hardness distribution of A387–SUS 405 and A387–SUS 405' after testing was similar, although the hardness of the base metals decreased with strain cycling on the whole.

**Strain Distribution**

The strain distribution at $N_f/2$ in A387–SUS 405 is shown in Fig. 12, the arrow mark denoting the rupture point. From Fig. 12 it is found that the strain concentration occurred on the SUS 405 side in a high strain range. On the other hand, it occurred on the heat affected zone in a low strain range. The region where the strain concentration occurred corresponded well with the rupture point. This tendency was also found in the other stage of strain cycling. The broken lines in Fig. 12 will be described later. In many cases, the surface cracks had already been initiated at $N_f/2$. The reason for the average value of strain measured along the gauge length in Fig. 12 being larger than the total strain range set in the experiment was that the change in spacing of the moiré grids contained not only the displacement due to plastic deformation but also the crack opening displacement due to these cracks initiated on the specimen surface.

The strain distribution at $N_f/2$ in A387–SUS 405' is shown in Fig. 13. The strain concentration occurred on the SUS 405' side. In A387–SUS 405', the region where the strain concentration occurred also

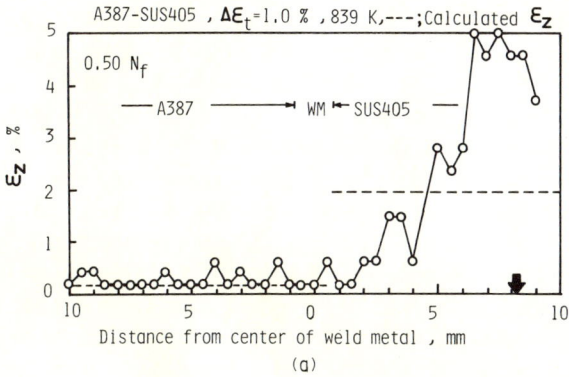

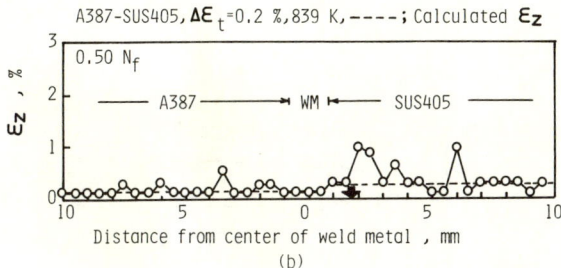

FIG. 12. Strain distribution in A387–SUS 405 at $N_f/2$: (a) $\Delta\varepsilon_t = 1.0\%$; (b) $\Delta\varepsilon_t = 0.2\%$.

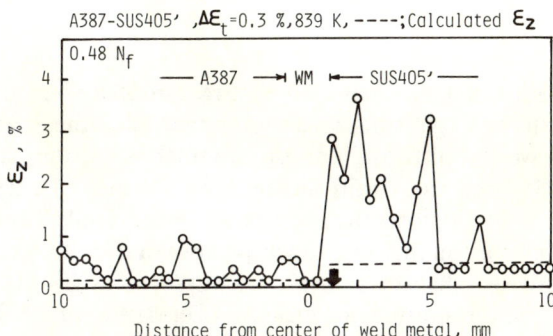

FIG. 13. Strain distribution in A387–SUS 405′ at $N_f/2$ ($\Delta\varepsilon_t = 0.3\%$).

corresponded well with the rupture point, this also being found in the other stage of strain cycling and in the other strain ranges.

The strain distribution of A387–SUS 316 varied with strain cycling, because A387 and SUS 316 cyclically softened and hardened, respectively. A typical variation of average strain on both base metal sides with strain cycling is shown in Fig. 14. As already described, the average value of the strain measured was not always equal to $\Delta\varepsilon_t$ set in the experiment due to the crack initiated. The points plotted in Fig. 14 were obtained by proportionally distributing the average strain measured in each base metal so that the average of the strain measured along the gauge length was equal to $\Delta\varepsilon_t$. Figure 14 indicates that the higher strain occurred on the SUS316 side at an early stage of strain cycling, and that the strain concentration occurred on the A387 side during the subsequent strain cycling.

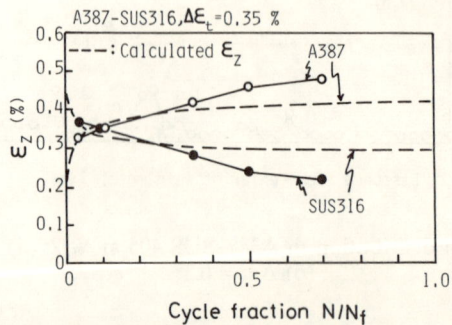

FIG. 14. Variation of average strain in each base metal with strain cycling in A387–SUS 316 ($\Delta\varepsilon_t = 0.35\%$).

## Characterization of Low Cycle Fatigue Fracture Behavior

From the foregoing results, it can be deduced that the fatigue lives of DMEBWJs were significantly shorter than those of the base metals. This resulted from the strain concentration which occurred in the material with lower deformation resistance (Figs. 12–14). There was also an inverse proportionality relationship between the strain distribution and the hardness distribution. For example, in A387–SUS 405 and A387–SUS405′, the strain concentration occurred on the SUS 405 and SUS 405′ sides, respectively, where the hardness was relatively low during all stages of strain cyclings (Figs. 2 and 4). While the strain distribution varied with strain cycling in A387–SUS 316, the number of cycles at which the strain concentration occurred on the A387 side corresponded with the number of cycles at which the hardness of A387 became lower than that of SUS 316 (Figs. 11 and 14). The number of cycles at which the higher strain was produced on the A387 side in A387–SUS 316 also corresponded to that at which the stress range was at its peak value (Figs. 7 and 14). Therefore, it can be concluded that the hardness distribution is an important aspect in which the strain distribution is reflected. It is worth noting that the information on the initial hardness distribution is not always adequate and it is necessary to determine the hardness variation with strain cycling.

In general, a DWEBWJ ruptured in the material with lower deformation resistance. Strictly speaking, not only mechanical factors such as hardness distribution and plastic constraint [13], but also metallurgical factors such as discontinuities in the microstructure and carbide precipitation near the bond region [8, 9], markedly affect the fracture of a DMEBWJ. For example, the fracture of A387–SUS 405 in the lower strain range would start with the strain concentration that occurs in the heat-affected zone with lower hardness, because the effect of plastic constraint is relatively little, so that A387–SUS 405 would rupture in the heat-affected zone. In the higher strain range, on the other hand, the effect of plastic constraint becomes notable and the deformation of the heat-affected zone with lower hardness is constrained. As the result, A387–SUS 405 would rupture on the base metal side. As an analogy, it is thought that A387–SUS 316 would rupture on the base metal side of A387, because of the plastic constraint. It is not yet clear why A387–SUS 405′ ruptured at the bond region on the SUS 405 side. Perhaps both plastic constraint near the bond region on the SUS 405′ side and carbide precipitation near the bond region [1, 8, 9] would affect the fracture point of A387–SUS 405′, and this needs further investigation.

## LIFE PREDICTION OF DMEBWJS

A simple method for life prediction of a DMEBWJ is proposed by representing the cyclic hardening and softening properties of the base metals as a function of the cumulative plastic strain. In this work, emphasis was put on the need for simplicity.

As a simple empirical equation which represents the cyclic hardening and softening behaviour of a material, the following equations have been proposed [14]:

$$\Delta\sigma = k(\Delta\varepsilon_p)^n \tag{1}$$

$$k = C\bar{\varepsilon}_p^m \tag{2}$$

where $\bar{\varepsilon}_p = \Sigma(\Delta\varepsilon_p)$ and $\Delta\sigma$, $\Delta\varepsilon_p$ and $\bar{\varepsilon}_p$ are the stress range, plastic strain range and cumulative plastic strain, respectively, and $n$, $m$ and $C$ are material constants. Hereafter, the suffices A3, S4, S4' and S3 are added to these material constants for A387, SUS 405, SUS 405' and SUS 316, respectively. The scatter of $n$ determined from the hysteresis loops for each base metal is summarized in Table 4. In this work, the value of $n$ was treated as a constant for simplicity, these constant values being given in Table 4. The relationships between $k$ and $\bar{\varepsilon}_p$ can be deduced from Figs. 15–18, which indicate that $k$ can be represented by a linear or bi-linear equation as a function of $\bar{\varepsilon}_p$. The values of $C$ and $m$ obtained by the least squares method are summarized in Table 4.

As a first approximation, it can be considered that the joints tested in this work consisted of two base metals, which are denoted by $q$ and $r$ ($q$, $r$ = A3, S4, S4' or S3), and these suffices are added to the material constants of $n$, $m$ and $C$. For example, $n_q$ denotes the $n$ value in Eqn. (1) for the base metal $q$. In this case, the following equation can be obtained about the stress condition, because the stress ranges $\Delta\sigma_i$ at cycle $i$ applied to the base metal are the same:

$$\Delta\sigma_i = k_q(\Delta\varepsilon_{p,q,i})^{n_q} = k_r(\Delta\varepsilon_{p,r,i})^{n_r} \tag{3}$$

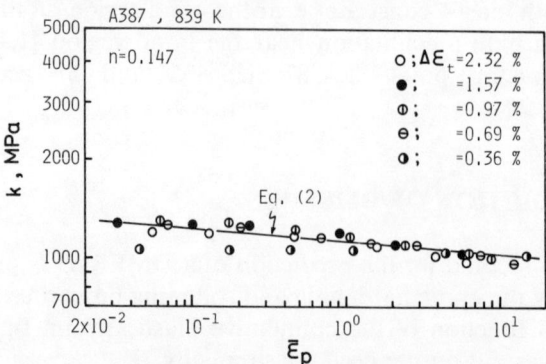

FIG. 15. Relationship between $k$ and $\bar{\varepsilon}_p$ in A387.

TABLE 4
Summary of material constants

| Material | Scatter of $n$ | $n$ value used | $C$ (MPa) | $m$ | $A$ | $B$ $(\times 10^{-3})$ | $\alpha$ | $\beta$ |
|---|---|---|---|---|---|---|---|---|
| A387 | 0·131–0·169 | 0·147 | 1127 | −0·040 | 1·077 | 5·940 | 0·737 | 0·110 |
| SUS 316 | 0·152–0·258 | 0·230 | 2340 ($\bar{\varepsilon}_p$ <0·66) 2250 ($\bar{\varepsilon}_p$ ≥0·66) | 0·095 ($\bar{\varepsilon}_p$ <0·66) 0 ($\bar{\varepsilon}_p$ ≥0·66) | 0·244 | 12·40 | 0·529 | 0·158 |
| SUS 405 | 0·062–0·087 | 0·071 | 507 ($\bar{\varepsilon}_p$ <0·25) 480 ($\bar{\varepsilon}_p$ ≥0·25) | −0·011 ($\bar{\varepsilon}_p$ <0·25) −0·047 ($\bar{\varepsilon}_p$ ≥0·25) | 0·539 | 3·217 | 0·605 | 0·071 |
| SUS 405' | 0·065–0·089 | 0·075 | 422 | −0·009 | 2·239 | 4·460 | 0·731 | 0·133 |

$\Delta\sigma = k(\Delta\varepsilon_p)^n$; $k = C\bar{\varepsilon}_p^m$; $\Delta\varepsilon = AN_f^{-\alpha} + BN_f^{-\beta}$.

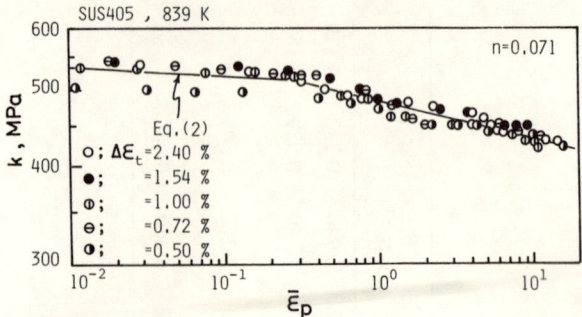

FIG. 16. Relationship between $k$ and $\bar{\varepsilon}_p$ in SUS 405.

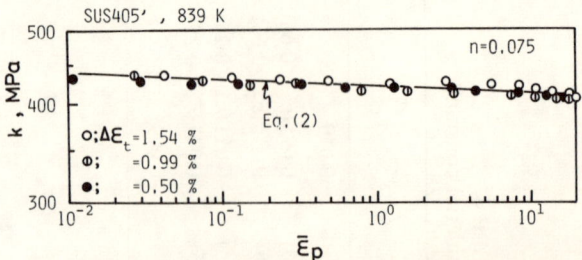

FIG. 17. Relationship between $k$ and $\bar{\varepsilon}_p$ in SUS 405'.

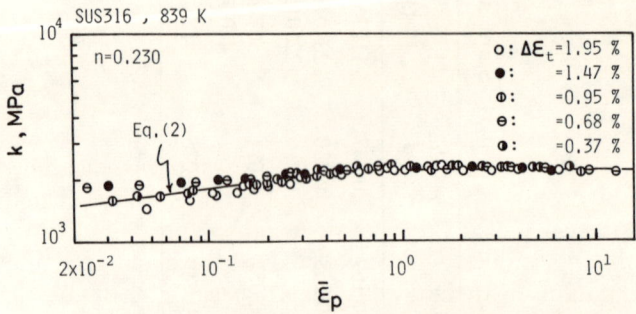

FIG. 18. Relationship between $k$ and $\bar{\varepsilon}_p$ in SUS 316.

where $\Delta\varepsilon_{p,q,i}$ and $\Delta\varepsilon_{p,r,i}$ are the plastic strain ranges at cycle $i$ on $q$ and $r$, respectively. The following equations are also satisfied for displacement, because the total displacement of a DMEBWJ is given by the summation of the displacements of each base metal:

$$\Delta\varepsilon_{q,i}L_q + \Delta\varepsilon_{r,i}L_r = \Delta\varepsilon_t(L_q + L_r) \qquad (4)$$

$$\Delta\varepsilon_{q,i} = \Delta\varepsilon_{p,q,i} + \Delta\sigma_i/E_q \tag{5}$$

$$\Delta\varepsilon_{r,i} = \Delta\varepsilon_{p,r,i} + \Delta\sigma_i/E_r \tag{6}$$

where $\Delta\varepsilon_{q,i}$ and $\Delta\varepsilon_{r,i}$ are the total strain ranges at cycle $i$ in the base metals $q$ and $r$, respectively, and $\Delta\varepsilon_t$ is the overall total strain range set in the experiment. $L_q$ and $L_r$ in Eqn. (4) are the respective lengths of base metals $q$ and $r$ occupying the gage length, and $E$ is Young's modulus. In this work, $E = 1.57 \times 10^5$ MPa was used for all base metals. Based on the strain distribution shown in Figs. 12–14, the deformation resistance of the weld metal in both A387–SUS 405 and A387–SUS 405′ was approximated to be equal to that of A387. The deformation resistance of the weld metal in A387–SUS 316 was approximated to be equal to that of SUS 316. According to this approximation, the values of $L_q$ and $L_r$ were as follows: $L_{A3} = 11$ mm, $L_{S4} = L_{S4'} = 9$ mm (in A387–SUS 405 and A387–SUS 405′) and $L_{S3} = 11$ mm, and $L_{A3} = 9$ mm (in A387–SUS 316). Substituting Eqns. (1) and (2) into Eqns. (3)–(6), the plastic and total strain ranges on the base metals $q$ and $r$ can be obtained by solving the simultaneous equations regarded as $\Delta\varepsilon_{p,q,i}$ and $\Delta\varepsilon_{p,r,i}$. The stress range at the $i$th cycle can be also obtained by substituting the calculated $\Delta\varepsilon_{p,q,i}$ and $\Delta\varepsilon_{p,r,i}$ values into Eqn. (3)

The stress range of each DMEBWJ with strain cycling was calculated by repeating this procedure for each cycle, and the results obtained are given in Figs. 6–8 by the broken lines, where the stress–strain relationship at the first cycle was represented as follows by applying Eqn. (3):

$$\Delta\sigma_1/2 = k(\Delta\varepsilon_{p,1}/2)^n$$

$$k = C\bar{\varepsilon}_p^m = C(\Delta\varepsilon_{p,1}/2)^m \tag{7}$$

From these figures it was found that the stress range of each DMEBWJ with strain cycling was well predicted by this simple calculation, although the calculated values of stress range were slightly lower than the experimental values. This would be due to neglecting the multiaxial stress condition near the weld metal, because a higher load is necessitated under a multiaxial stress condition than under a uniaxial condition in order to produce the same axial strain in the strain-controlled test [4, 13].

The broken lines in Figs. 12–14 indicate the average strain on each base metal in the joint calculated by the above procedure, the average strain on each base metal calculated being in good agreement with the experimental results. It is worth noting that the variation of strain with strain cycling on both base metals in A387–SUS 316 was well predicted.

Provided that the material fails when the accumulated damage reaches unity [15], the fatigue life of a DMEBWJ can be predicted by the following procedure.

(i) Calculate the damage to each base metal for each cycle (inverse of the fatigue life) by substituting the $\Delta\varepsilon_{q,i}$ and $\Delta\varepsilon_{r,i}$ values obtained from the new procedure into the fatigue life equation of the base metals, which are represented by the following equation:

$$\Delta\varepsilon = AN_f^{-\alpha} + BN_f^{-\beta} \tag{8}$$

where $A$, $B$, $\alpha$ and $\beta$ are material constants. The values of these material constants are summarized in Table 4.

(ii) Repeat (i) for each cycle until the cumulative damage to each base metal reaches unity, and then obtain the fatigue lives of the base metals.

(iii) Compare the fatigue lives of both base metals, and the shorter fatigue life gives the fatigue life of the DMEBWJ. For example, when the cumulative damage of A387 reaches unity first, the DMEBWJ will rupture on the A387 side.

The fatigue lives predicted by this procedure are compared with the experimental values in Fig. 19, from which it can be seen that the predicted values are nearly consistent with the experimental values. Furthermore, the fracture points were estimated to be on the SUS 405

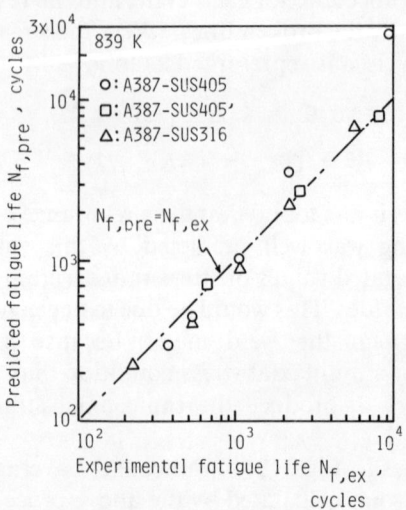

FIG. 19. Comparison between the predicted fatigue lives and experimental results.

side in A387–SUS 405, the SUS 405' side in A387–SUS 405', and the A387 side in A387–SUS 316, which also agree with the experimental results. However, the prediction gives a longer fatigue life in a lower strain range in A387–SUS 405, this resulting mainly from the neglect of strain concentration which occurred in the heat-affected zone with lower hardness shown in Fig. 12(b). Although the effect of plastic constraint on the fatigue life of a DMEBWJ was neglected in this work, the proposed method gave a comparatively good prediction result, not only for A387–SUS 316 being ruptured on the base metal side, but also for A387–SUS 405' being ruptured at the bond region. This suggests that the effect of plastic constraint near the bond region on the fatigue life of a DMEBWJ is smaller than would be expected.

## CONCLUSIONS

The low cycle fatigue strength of DMEBWJs was investigated. Using specimens on whose surface a heat-resisting moiré grid had been inscribed, the local strain distribution in each joint was measured, and the relationship between the strain distribution and fatigue fracture behavior was discussed. Moreover, a simple method for predicting the fatigue life of a DMEBWJ was proposed.

It was found that the low cycle fatigue strength of a DMEBWJ was notably shorter than those of its base metals. This was due to the strain concentration which occurred on the side of the material with the lower deformation resistance. It was also found that the hardness distribution was an important measure in which the strain distribution was reflected. In this case, information only on the initial hardness distribution was not always adequate, and it was necessary to ascertain the hardness variation with strain cycling. For example, the strain distribution in A387–SUS 316 varied with strain cycling, and the number of cycles at which the strain concentration occurred on A387 corresponded with that at which the hardness of A387 became lower than that of SUS 316. A simple fatigue life prediction method for a DMEBWJ is proposed by representing the cyclic softening and hardening properties of the base metals. The estimated variation of average strains on the base metals with strain cycling were well consistent with the experimental results. The predicted fatigue lives of the joints were in good agreement with the experimental results, except for the case in which the strain concentration occurred on a heat-affected zone with lower hardness than the base metals.

## ACKNOWLEDGMENT

A Grant in Aid for Scientific Research by the Ministry of Education of Japan is gratefully acknowledged.

## REFERENCES

[1]   R. L. Klueh, J. F. King and L. F. Griffith, *Weld. J.,* **62**, 154-s (1983).
[2]   C. D. Lundin, *Weld. J.,* **61**, 58-S (1982).
[3]   S. Shimizu and Y. Ikemoto, *Trans. Jpn. Soc. Mater. Sci.,* **28**, 441 (1979).
[4]   I. Nonaka, M. Kitagawa and A. Ohtomo, *Trans. Jpn. Soc. Mater. Sci.,* **31**, 578 (1982).
[5]   J. Fukakura and T. Mori, *J. Test. Eval.,* **14**, 7 (1986).
[6]   M. Okazaki, Y. Mutoh, M. Tabata, K. Hatakeyama and Y. Itoh, *Trans. Jpn. Soc. Mater. Sci.,* **36**, 280 (1987).
[7]   M. Okazaki, Y. Mutoh, M. Yamaguchi and Y. Itoh, *Trans. Jpn. Soc. Mater. Sci.,* **36**, 286 (1987).
[8]   R. D. Nicholson, *Met. Technol.,* **9**, 305 (1982).
[9]   R. Viswanathan, *Trans. ASME, J. Press. Vess. Technol.,* **107**, 218 (1985).
[10]  H. Shimada and M. Obata, *J. Jpn. Soc. Non-Destructive Inspection,* **22**, 627 (1973).
[11]  T. Fujita, *Heat Treatment of Stainless Steel,* Nikkan Kogyo Shinbun-sya, Chapters 3 and 4 (1970).
[12]  For example, G. R. Halford, M. H. Hirschberg and S. S. Manson, Fatigue at Elevated Temperatures, *ASTM STP 520,* 658, (1973).
[13]  M. Toyoda and K. Satoh, *Trans. Jpn. Weld. Soc.,* **37**, 58 (1968).
[14]  J. M. Corum, *ORNL Report 5301,* 56 (1977).
[15]  M. A. Miner, *Trans. ASME, J. Appl. Mech.,* **67**, 1599 (1945).

# Thermal–Mechanical Fatigue Failure and Life Prediction

AKITO NITTA AND KAZUO KUWABARA

*Nuclear Engineering Department, Komae Research Laboratory, Central Research Institute of Electric Power Industry, 2-11-1, Iwato-kita, Komae-shi, Tokyo 201, Japan*

## ABSTRACT

Failure modes and life prediction method are presented for thermal–mechanical low cycle fatigue (TMF) in metallic materials for high temperature applications. By characterizing TMF failure under in-phase and out-of-phase strain–temperature cycling conditions for several kinds of steels and alloys, the life relationship between both cycling conditions was classified into four types associated with environmental and creep effects. In general, excepting for a case showing the large effect of creep, the TMF life could be conservatively predicted by applying the relationship based on total strain range that was obtained from other mechanical properties, particularly the total strain range versus life relationship in isothermal fatigue at the maximum temperature of TMF. The strain range partitioning method was also applicable for predicting the TMF life of ductile steels, but not for inductile superalloys whose TMF life was markedly dependent on the elastic strain or stress. Using a new approach in which the high temperature low cycle fatigue life was correlated with *strain energy parameters* derived from the macrocrack propagation laws, the TMF life could be well predicted from the characteristics of isothermal fatigue failure and crack propagation for all the materials quoted, including superalloys.

## INTRODUCTION

Thermal fatigue in the low cycle range (less than $10^4$–$10^5$ cycles) is one of the damage factors to be considered in the design and operation of high temperature equipment. For instance, the operation of a steam turbine rotor is controlled by using the *low cycle fatigue index* (LCFI) to limit the fatigue damage during transient operations such as start-up and shut-down. In older fossil-fuel power plants whose service life is being extended beyond their design life, these transient operations have been frequently repeated for load control over a number of years. This

203

requires an accurate assessment of the thermal fatigue damage accumulated in the boiler and turbine components of the power plant. Similarly, an assessment methodology for thermal fatigue damage is essential for the design of high temperature components.

Thermal fatigue, which is a failure phenomenon caused by constraints on free thermal expansion, can be grouped into two categories, i.e. *thermal–stress fatigue* (TSF) due to internal constraints, and *thermal–mechanical fatigue* (TMF) due to external constraints [1]. A TSF failure can be experimentally characterized by using thermal cycling test equipment (e.g. a Glenny-type apparatus with hot and cold fluidized beds [2]) with specimens simulating the shape of actual components. This laboratory test is effective for a material-to-material comparison of TSF strength [3]. In this case, however, the stress and strain conditions in the specimen tested must be analytically obtained. On the other hand, a modern servo-controlled fatigue machine developed from a Coffin-type apparatus [4] is usually applied to characterize a TMF failure. This test is more complicated than the TSF test, but it has the advantage of independently controlling the strain and temperature cycles on a uniaxial specimen under predetermined conditions. Therefore, it can be said that this type of test is suitable for investigating the mechanical laws of thermal fatigue failure.

For the last decade, the authors have been experimentally studying TMF failure in Cr–Mo steels, stainless steels and superalloys by using computer-controlled electro-hydraulic fatigue testing machines [5, 6]. In this chapter, which summarizes the experimental results obtained to date, the TMF failure is characterized and correlated with other mechanical properties to predict the TMF life. The chapter also describes the methodology for TMF life prediction.

## CHARACTERIZATION OF THERMAL–MECHANICAL FATIGUE FAILURE

### Life Relationships between In-phase and Out-of-phase Cyclings

The TMF tests are usually conducted with in-phase and out-of-phase cyclings, as shown schematically in Fig. 1, which are the extremes of the test conditions. From the life $N_f$ relationship between both cyclings, the TMF life characteristics can be classified into the following four types:

Type I, $N_f$ for in-phase cycling is shorter in a low strain range;
Type O, $N_f$ for out-of-phase cycling is shorter in a low strain range;

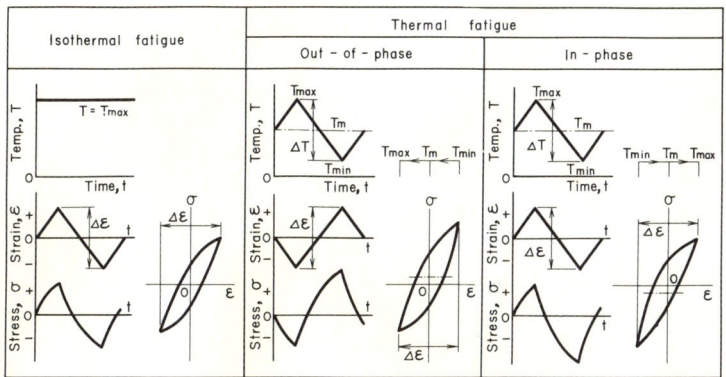

FIG. 1. Schematic diagram showing the wave forms for temperature, strain and stress in thermal–mechanical and isothermal fatigue tests.

Type E, $N_f$ is nearly equal for in-phase and out-of-phase cycling;
Type E', $N_f$ for in-phase cycling is shorter in a high strain range, but nearly equal to that for out-of-phase cycling in a low strain range.

Typical examples of these types are shown in Fig. 2 as a relationship between the inelastic strain range $\Delta\varepsilon_{in}$ and $N_f$. A similar classification can also be applicable to the relationship between the total strain range $\Delta\varepsilon$ and $N_f$.

Figure 3 shows the TMF lives of various materials for $\Delta\varepsilon = 1.5\%$ and $0.5\%$ at frequencies of $0.003$ to $0.017$ Hz. In order to illustrate the difference in $N_f$ according to the material and test temperature, other published data [7, 8] are cited in this figure. Except for the superalloys, the TMF lives of various materials in their practical temperature ranges are almost equal, i.e. about 500 and 5000 cycles at $\Delta\varepsilon = 1.5\%$ and $0.5\%$, respectively. In this case, the TMF life characteristics are of Type O or Type E, as can be seen in Fig. 3. In the $\Delta\varepsilon_{in}$ versus $N_f$ relationship, the lives at $\Delta\varepsilon_{in} = 1.0\%$ and $0.1\%$ are also about 500 and 5000 cycles, respectively. On the other hand, the superalloys have a shorter life than other ductile materials, but the life in a low strain range tends to approach that of other materials.

**Correlation between TMF Life and Failure Mode**
As can be seen from Fig. 3, in which the same classification is also indicated, the TMF life characteristics change from Type O or Type E to Type I with an increase in the maximum temperature $T_{max}$ for a given material (e.g. Cr–Mo–V forged steel and 304 SS). In this case, the failure

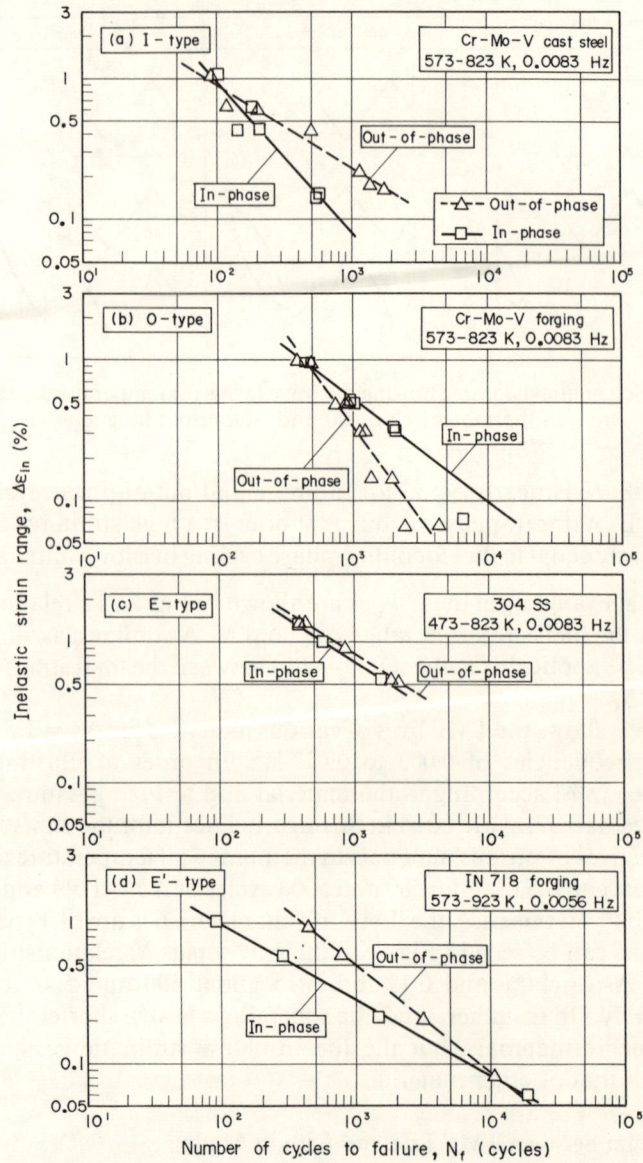

FIG. 2. Typical examples showing four types of TMF life characteristics for the inelastic strain range versus failure life relationship.

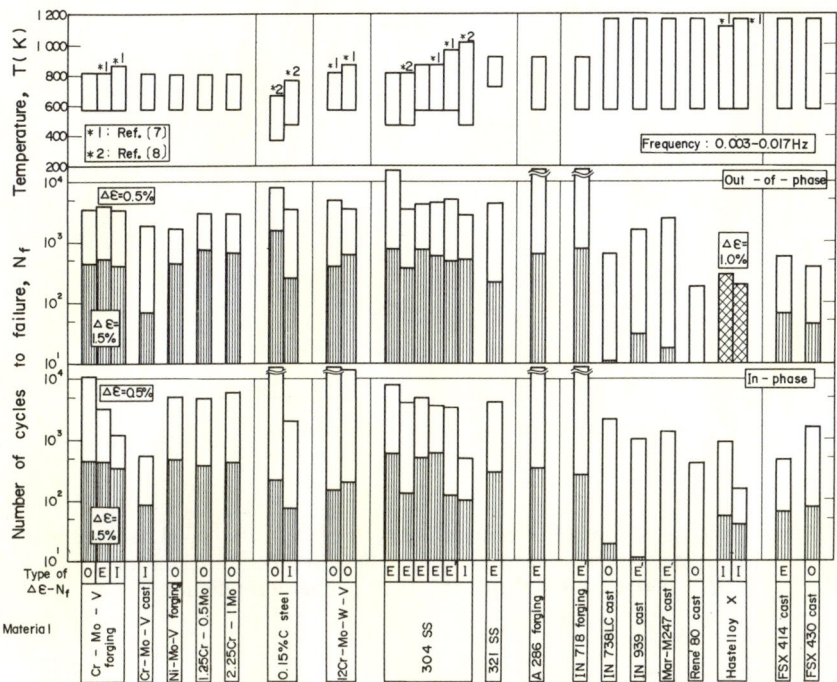

FIG. 3. TMF lives at total strain ranges of 0·5% and 1·5% for 18 kinds of materials.

mode for in-phase cycling was of the intergranular type, while striations could be seen on the fracture surface for out-of-phase cycling. This means that Type I failure was caused by the creep effect. Such a reduction in $N_f$ due to the creep effect was also observed in the practical temperature range of a material (e.g. Cr–Mo–V forged steel [9] and 304 SS [10, 11]) when holding the tensile strain at $T_{max}$ for in-phase cycling. Figure 4 shows an example of the hold-time effect on the TMF life of 304 SS. In this case, $N_f$ for in-phase cycling decreased with an increase in the proportion of intergranular cracks [11].

Type O failure occurs mainly in Cr–Mo steels. This can be interpreted by a difference in the crack initiation life $N_i$ between in-phase and out-of-phase cyclings in a low strain range as shown in Fig. 5. The difference of $N_i$ (i.e. earlier crack initiation for out-of-phase cycling) was related to the cracking behavior of the surface oxide scale [12]. As shown in Fig. 6, for out-of-phase cycling with high tensile stress at low temperature (see

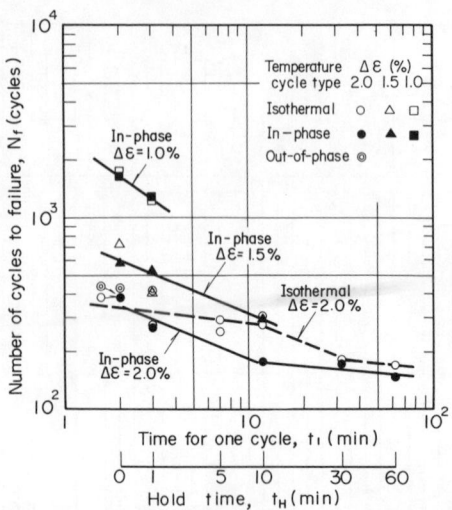

FIG. 4. Effect of strain hold time at 550 °C on the failure life of 304 stainless steel in thermal–mechanical and isothermal fatigue (temperature conditions: 200–550 °C in TMF and 550 °C in isothermal fatigue).

Fig. 1), the surface oxide scale cracked much more easily than for in-phase cycling, and the oxide cracks then accelerated the nucleation of a fatigue crack in the underlying metal. In both the cyclings, however, the failure mode was of the transgranular type. In other words, a Type O failure, at least in Cr–Mo steels, was due to the environmental effect. A similar effect was observed with isothermal low cycle fatigue in air for a $2\frac{1}{4}$Cr–1Mo steel [13, 14]. In this case, the isothermal fatigue life with a compressive strain hold time was shorter than that with a tensile strain hold time in a low strain range. A Cr–Mo–V cast steel in Fig. 3 shows Type I as an exception. This was caused by the acceleration of fatigue crack initiation due to internal casting defects for in-phase cycling, but not due to the creep effect.

Type E, in which a failure of the transgranular type occurred with both the cyclings, is regarded as a case with little effect from creep and/or the environment. In the case of Type E' that has been observed in some superalloys such as IN 718, the failure was of the intergranular mode implying that the creep effect occurred for in-phase cycling, while the failure mode for out-of-phase cycling was of the transgranular type [15].

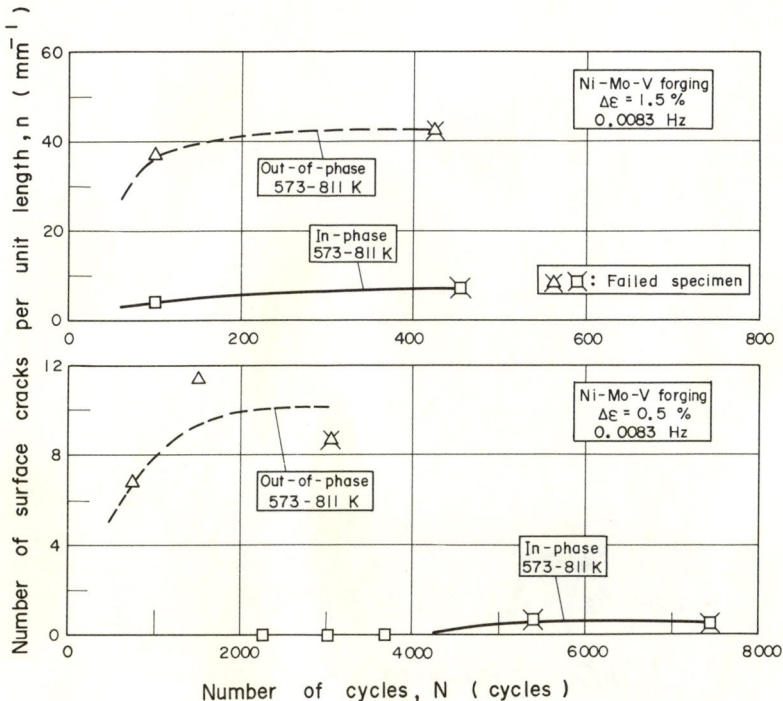

FIG. 5. Change in number of surface cracks with number of cycles in an Ni-Mo-V forged steel showing TMF life characteristics of Type O.


## THERMAL-MECHANICAL FATIGUE LIFE PREDICTION

### Life Prediction from Other Mechanical Properties

To predict the TMF life from correlations with other mechanical properties, two approaches (i.e. $\Delta\varepsilon$- and $\Delta\varepsilon_{in}$-based ones) are available.

With the $\Delta\varepsilon$-based approach, the $\Delta\varepsilon$ versus $N_f$ relationship in isothermal fatigue at $T_{max}$ can be applied. The universal slope and 10% rule method is also applicable to the $\Delta\varepsilon$-based life prediction. The $\Delta\varepsilon$ versus $N_f$ relationship in the universal slope method can be expressed as follows [16]:

$$\Delta\varepsilon = \frac{3 \cdot 5\sigma_B}{E} N_f^{-0.12} + D^{0.6} N_f^{-0.6} \tag{1}$$

where $\sigma_B$, $E$ and $D$ are the ultimate strength, Young's modulus and

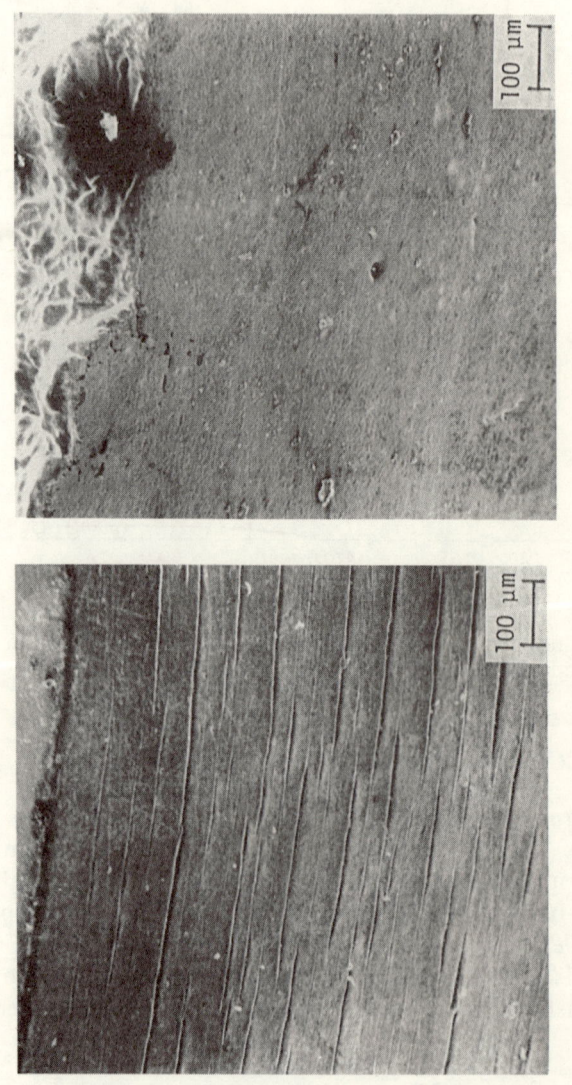

(b)

(a)

FIG. 6. Scanning electron micrographs showing the typical surface appearance of failed TMF specimens in a low strain range for low alloy steels (Ni–Mo–V forged steel, $\Delta\varepsilon = 0.5\%$; stress axis in the vertical direction). (a) Out-of-phase; (b) in-phase.

tensile fracture ductility at room temperature, respectively. The 10% rule gives one-tenth of $N_f$ obtained from Eqn. (1) as the life at high temperature [16].

With the $\Delta\varepsilon_{in}$-based approach the $\Delta\varepsilon_{in}$ versus $N_f$ relationship in isothermal fatigue at $T_{max}$ as well as the universal life relationships in the strain range partitioning method are applicable. The universal life relationships are expressed by the following equation [17]:

$$\Delta\varepsilon_{ij}/D_i = A_{ij}N_{ij}^{-\alpha_{ij}} \qquad (i, j = \text{p or c}) \qquad (2)$$

where $\Delta\varepsilon_{ij}$ is the partitioned inelastic strain range, $D_i$ is the tensile ($i = $ p) or creep ($i = $ c) ductility at $T_{max}$, and $A_{ij}$ and $\alpha_{ij}$ are constants. However, for expressing the time-independent (i.e. purely cycle-dependent) $\Delta\varepsilon_{in}/D_p$ versus $N_f$ relationship, the following equation may be conservative rather than the original PP-type relationship [5]:

$$\Delta\varepsilon_{in}/D_p = 1\cdot25N_f^{-0\cdot8} \qquad (3)$$

Table 1 indicates the predicted TMF life by these two approaches. In general, the $\Delta\varepsilon$-based prediction is better than the $\Delta\varepsilon_{in}$-based approach, and the $\Delta\varepsilon$ versus $N_f$ relationship obtained isothermally at $T_{max}$ provides the best results. The universal slope method combined with the 10% rule is the second best. The $\Delta\varepsilon_{in}$ versus $N_f$ relationship in isothermal fatigue at $T_{max}$ gives relatively good results for such ductile materials as Cr–Mo and stainless steels, but it is generally dangerous to use it for inductile materials such as superalloys, which are prone to elastic fatigue even in the low cycle range.

### Application of the Strain Range Partitioning Method

The strain range partitioning method [17, 18] is effective for predicting the TMF life of ductile materials [9], but is not applicable for inductile materials whose inelastic strain is low because of their high strength [5]. To overcome this problem, Halford and Saltsman [19] have recently proposed a total strain range version in which the effect of elastic strain is taken into account.

To apply the strain range partitioning method to life prediction, the inelastic strain must be partitioned into the plastic and creep strains. In other words, this life prediction method requires an analysis of the hysteresis loop for the fatigue cycle in question. In the case of TMF, it is

TABLE 1

Results of TMF life prediction based on the correlation with other mechanical properties

| Material | Temperature range $T_{min}$–$T_{max}$ (K) | Homologous temperature $T_{max}/T_m$ | Frequency ν (Hz) | Cyclic deformation | Type[1] of $\Delta\varepsilon$–$N_f$ | $\Delta\varepsilon$-based prediction[2] Universal slope + 10% rule[3] OUT | IN | Isothermal $N_f$ at $T_{max}$ OUT | IN | Type[1] of $\Delta\varepsilon_{in}$–$N_f$ | $\Delta\varepsilon_{in}$-based prediction[2] PC-type universal slope[4] OUT | IN | Isothermal $N_f$ at $T_{max}$ OUT | IN | Reference |
|---|---|---|---|---|---|---|---|---|---|---|---|---|---|---|---|
| Cr–Mo–V forging | 573–823 | 0·46 | 0·008 3 | Soft | O | ○ | ○ | ○ | ○ | O | × | ○ | ○ | ○ | This work |
| | 573–823 | 0·46 | 0·008 3 | Soft | E | ◉ | ◉ | ○ | ○ | E | × | × | ○ | ○ | JSMS [7] |
| | 573–873 | 0·49 | (7–8·3) × 10⁻³ | Soft | I | ◉ | ◉ | ○ | ○ | I | ○ | × | ○ | × | JSMS [7] |
| Cr–Mo–V cast steel | 573–823 | 0·46 | 0·008 3 | Soft | I | × | × | × | × | — | ○ | × | ○ | ○ | This work |
| Ni–Mo–V forging | 573–811 | 0·46 | 0·008 3 | Soft | O | ○ | ○ | ◉ | ◉ | O | × | ○ | ○ | ○ | This work |
| 1·25Cr–0·5Mo steel | 573–811 | 0·45 | 0·008 3 | Soft | O | ◉ | ◉ | ○ | ○ | O | ○ | ○ | ○ | ○ | This work |
| 2·25Cr–1Mo steel | 573–811 | 0·45 | 0·008 3 | Soft | O | ◉ | ◉ | × | × | O | ◉ | ◉ | ○ | ○ | This work |
| 0·15%C steel | 373–673 | 0·37 | 0·016 7 | Hard | O | ○ | ○ | ○ | ○ | O | ◉ | ◉ | × | × | Fujino [8] |
| | 473–773 | 0·43 | 0·016 7 | Hard | I | ○ | ○ | × | × | I | × | × | × | × | Fujino [8] |
| 12Cr–Mo–W–V steel | 573–823 | 0·47 | 0·008 3 | Soft | O | ◉ | ◉ | ○ | ○ | O | × | ○ | ○ | ○ | JSMS [7] |
| | 573–873 | 0·49 | 0·008 3 | Soft | O | ○ | ○ | ○ | ○ | O | × | × | ○ | ○ | JSMS [7] |
| 304 SS | 473–823 | 0·48 | 0·008 3 | Hard | E | ◉ | ◉ | ○ | ○ | E' | ○ | ○ | ○ | ○ | This work |
| | 473–823 | 0·48 | 0·008 3 | Hard | E | ◉ | ◉ | ○ | ○ | E' | × | × | ○ | ○ | Fujino [8] |
| | 573–873 | 0·51 | 0·008 3 | Hard | E | ○ | ○ | ○ | ○ | E' | ○ | ○ | ○ | ○ | This work |
| | 573–873 | 0·51 | 0·008 3 | Hard | E' | ○ | ○ | ○ | ○ | E' | ○ | ○ | ○ | ○ | JSMS [7] |
| | 573–973 | 0·57 | 0·005 6 | Hard | E' | ○ | ○ | × | × | I | ○ | ○ | × | × | JSMS [7] |
| | 473–1 023 | 0·60 | 0·008 3 | Hard | I | × | × | × | × | I | × | × | × | × | Fujino [8] |
| 321 SS | 723–923 | 0·56 | 0·003 3 | Hard | E | ○ | ○ | ○ | ○ | E | × | × | × | × | This work |

| Material | Temperature | | ductility | Condition | | | | | | | | | | | | Reference |
|---|---|---|---|---|---|---|---|---|---|---|---|---|---|---|---|---|
| A 286 forging | 573–923 | 0·005 6 | 0·56 | Soft | E | ⊚ | ⊚ | ⊚ | ⊚ | E | ⊚ | ⊚ | ⊚ | ⊚ | ○ | Q | This work |
| IN 718 forging | 573–923 | 0·005 6 | 0·6 | Soft | E' | ⊚ | ○ | ○ | ○ | E' | ○ | ○ | ○ | ⊚ | ○ | ○ | This work |
| IN 738LC cast | 573–1 173 | 0·005 6 | 0·75 | Hard | O | × | × | × | × | I | ○ | ○ | × | × | × | × | This work |
| IN 939 cast | 573–1 173 | 0·005 6 | 0·75 | Hard | E' | ○ | × | ○ | ○ | E' | ○ | ○ | × | × | × | × | This work |
| Mar-M247 cast | 573–1 173 | 0·005 6 | 0·75 | Hard | O | × | ○ | × | ○ | E' | ○ | ○ | × | × | × | × | This work |
| René 80 cast | 573–1 023 | 0·008 3 | 0·65 | Hard | I | × | × | × | × | I | × | × | × | × | × | × | JSMS [7] |
| Hastelloy X | 573–1 173 | 0·004 8 | 0·75 | Hard | I | × | × | × | × | I | × | × | × | ○ | ○ | × | JSMS [7] |
| FSX 414 cast | 573–1 173 | 0·005 6 | 0·73 | Hard | E | × | × | × | × | I | × | × | × | × | × | × | This work |
| FSX 430 cast | 573–1 173 | 0·005 6 | 0·73 | Hard | O | × | ○ | ○ | ⊚ | O | ○ | × | ⊚ | ⊚ | × | × | This work |

1. Diagrams (Δε, Δε_in vs $N_f$): Type I, Type O, Type E, Type E'

OUT : out-of-phase  
IN : in-phase

2. ⊚, conservative prediction; ○, good prediction within a factor of 1·5; ×, dangerous prediction.

3. $\Delta\varepsilon = (3\cdot5\sigma_B/E)(10N_f)^{-0\cdot12} + D^{0\cdot6}(10N_f)^{-0\cdot6}$, where $\sigma_B$ is tensile strength at room temperature, $E$ is Young's modulus at room temperature and $D$ is tensile ductility at room temperature.

4. $\Delta\varepsilon_{in}/D_p = 1\cdot25N_f^{-0\cdot8}$, where $D_p$ is tensile ductility at $T_{max}$.

not easy to partition the inelastic strain in an experimental way, although Halford and Manson [20] proposed the step-stress method as applicable to TMF. The step-stress method, however, appears to be complicated and difficult in practical use. Thus, the authors [21] have proposed two simplified methods, i.e. the *simplified step-stress method* and the *loop reversion method*.

In the original step-stress method, the creep strain was obtained as an integrated value of the steady-state creep rates determined by holding the stress at several levels on the hysteresis loop. According to the experimental findings of Manson *et al.* [22], the monotonic creep property was applicable to determine the steady-state creep rate. Thus, in the simplified method, the creep strain was evaluated as the area of the hatched triangle shown in Fig. 7 by using the monotonic creep property at $T_{max}$. In this case the creep strain was also assumed to be generated only during a heating period over the specified temperature of $T_c$ (e.g. $T_c = 500\,°C$ for low alloy steels and $550\,°C$ for austenitic stainless steels). The creep strain $\varepsilon_c$ can be expressed as follows:

$$\varepsilon_c = \frac{B\sigma_{T_{max}}^n t_1 (T_{max} - T_c)}{4(T_{max} - T_{min})} \qquad (4)$$

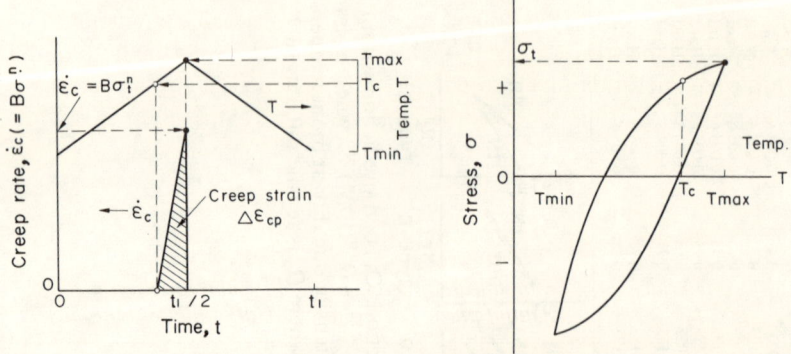

FIG. 7. Schematic diagram showing the simplified step-stress method for evaluating the creep strain in TMF under in-phase cycling.

where $\sigma_{T_{max}}$ is the stress at $T_{max}$, $t_1$ is time for one cycle without a hold time, $T_{min}$ is the minimum temperature, and $B$ and $n$ are constants in the following creep strain rate $\dot{\varepsilon}_c$ versus stress $\sigma$ relationship:

$$\dot{\varepsilon}_c = B\sigma^n \qquad (5)$$

Similarly, in the loop reversion method, the creep strain was assumed to be generated only during a heating period. The creep strain was partitioned in the same manner as the rapid straining method with the loop reversion as shown schematically in Fig. 8.

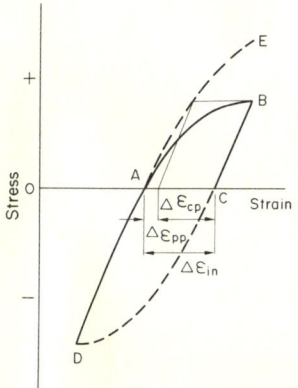

FIG. 8. Schematic diagram showing the loop reversion method for evaluating the creep strain in TMF under in-phase cycling.

Table 2 summarizes the results of a TMF life prediction in the range of strain rates from $4 \times 10^{-5}\,\mathrm{s}^{-1}$ to $3 \times 10^{-4}\,\mathrm{s}^{-1}$ for ductile materials by applying the strain range partitioning approach with the help of the simplified methods just quoted. The simplified step-stress method is relatively dangerous to use in comparison with the loop reversion

TABLE 2

Results of TMF life prediction by the strain range partitioning approach for ductile materials at a strain rate of $4 \times 10^{-5}$–$3 \times 10^{-4}\,\mathrm{s}^{-1}$

| Material | Simplified step-stress method | | Loop reversion method | |
|---|---|---|---|---|
| | Out-of-phase | In-phase | Out-of-phase | In-phase |
| Cr–Mo–V steel forging | ✕ | ○ | ○ | ◎ |
| Ni–Mo–V steel forging | ✕ | ○ | ○ | ◎ |
| $1\frac{1}{4}$Cr–$\frac{1}{2}$Mo–$\frac{3}{4}$Si steel | ○ | ○ | ○ | ○ |
| $2\frac{1}{4}$Cr–1Mo steel | ○ | ○ | ○ | ○ |
| 304 stainless steel | ○ | ○ | ○ | ◎ |
| 321 stainless steel | ✕ | ✕ | ○ | ○ |

✕, dangerous prediction; ○, good prediction within a factor of 2; ◎, conservative prediction.

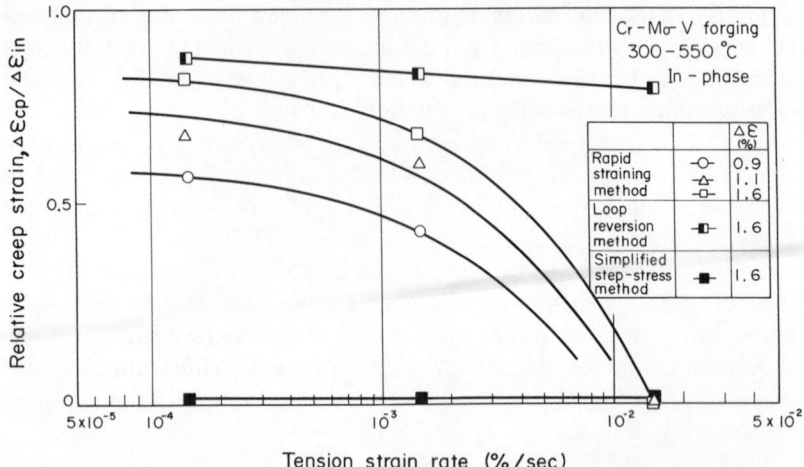

FIG. 9. Comparison between the creep strain evaluated by the simplified methods and the rapid straining method.

method, which is generally conservative. This tendency depends on the magnitude of the creep strain estimated by the two methods. In Fig. 9, showing an example of a Cr–Mo–V forged steel, the two simplified methods are compared with the accurate creep strain obtained by rapid straining for in-phase cycling. The creep strain was underestimated by the simplified step-stress method, while it was overestimated by the loop reversion method. However, the loop reversion method is available at low strain rates.

### New Approach Based on 'Strain Energy Parameters'

TMF crack initiation occurs on the surface of a specimen at an early stage of the strain cycles (see Fig. 5), and subsequent crack propagation causes the failure. Since the propagation rates for microcracks of a length of more than a few grain diameters can be extrapolated from the propagation laws of macrocracks (more than about 1 mm), the range in which the macrocrack propagation laws are applicable may be about 70% of $N_f$ [23]. Thus, new life relationships can be derived from the fracture mechanics laws on macrocrack propagation.

The crack propagation rates under isothermal low cycle fatigue conditions can be expressed as follows.

(i) Cycle-dependent (P-type) fatigue

$$dl/dN = C_f \Delta J_f^{m_f} \qquad (6)$$

(ii) Time-dependent (C-type) fatigue

or

$$dl/dt = C_c J^{*m_c} \tag{7a}$$

$$dl/dN = C_c \Delta J_c \qquad m_c = 1 \tag{7b}$$

where $\Delta J_f$ is the fatigue $J$-integral range, $J^*$ is the creep $J$-integral, $\Delta J_c$ is the creep $J$-integral range obtained by the time-integration of $J^*$ over a cycle, and $C_f$, $m_f$, $C_c$ and $m_c$ are material constants. As an example, the crack propagation characteristics of 304 SS are shown in Fig. 10. The relationships expressed by Eqns. (6) and (7) are almost independent of the temperature [23], and the characteristics of TMF crack propagation agree well with those under isothermal conditions (see Fig. 10) [24].

Assuming that $N_f$ is equal to the crack propagation life from the initial crack length $l_0$ to the final crack length $l_f$, the failure life laws can be obtained as follows [23].

(i) Cycle-dependent (P-type) fatigue

$$\Delta \tilde{W}_f^{m_f} N_f = D_f \tag{8}$$

$$\Delta \tilde{W}_f = \frac{\Delta\sigma\,\Delta\varepsilon_e}{2} + \frac{n'+1}{2\pi} f(n') \frac{\Delta\sigma\,\Delta\varepsilon_p}{n'+1} \tag{8a}$$

$$D_f = \frac{\ln(l_f/l_0)}{C_f 2\pi M_J} \qquad m_f = 1$$

$$D_f = \frac{l_0^{1-m_f} - l_f^{1-m_f}}{C_f(m_f-1)(2\pi M_J)^{m_f}} \qquad m_f \neq 1 \tag{8b}$$

(ii) Time-dependent (C-type) fatigue

$$\Delta \tilde{W}_c^{m_c} N_f = D_c t_0^{1-m_c} \tag{9}$$

$$\Delta \tilde{W}_c = \left[ \int_{-1}^{1} \left\{ \frac{1}{2^{2/3}} \frac{\sigma_{max}}{\Delta\sigma} \frac{\sigma_{max}\,\Delta\varepsilon_e}{2} + \frac{n+2}{2^{(n+1)/2+5/3}} (1+x)^{(n+1)/2} \right.\right.$$

$$\left.\left. \times \frac{n+1}{2\pi} f(n) \frac{\sigma_{max}\,\Delta\varepsilon_c}{n+1} \right\}^{3/2} dx \right]^{2/3} \qquad m_c \neq 1 \tag{9a}$$

$$\Delta \tilde{W}_c = \frac{\sigma_{max}}{\Delta\sigma} \frac{\sigma_{max}\,\Delta\varepsilon_e}{2} + \frac{n+2}{n+3} \frac{n+1}{2\pi} f(n) \frac{\sigma_{max}\,\Delta\varepsilon_c}{n+1} \qquad m_c = 1, \; \beta = 0.5 \tag{9b}$$

$$D_c = \frac{\ln(l_f/l_0)}{C_c 2\pi M_J} \qquad m_c = 1 \tag{9c}$$

$$D_c = \frac{l_0^{1-m_c} - l_f^{1-m_c}}{C_c(m_c-1)(2\pi M_J)^{m_c}} \qquad m_c \neq 1 \tag{9d}$$

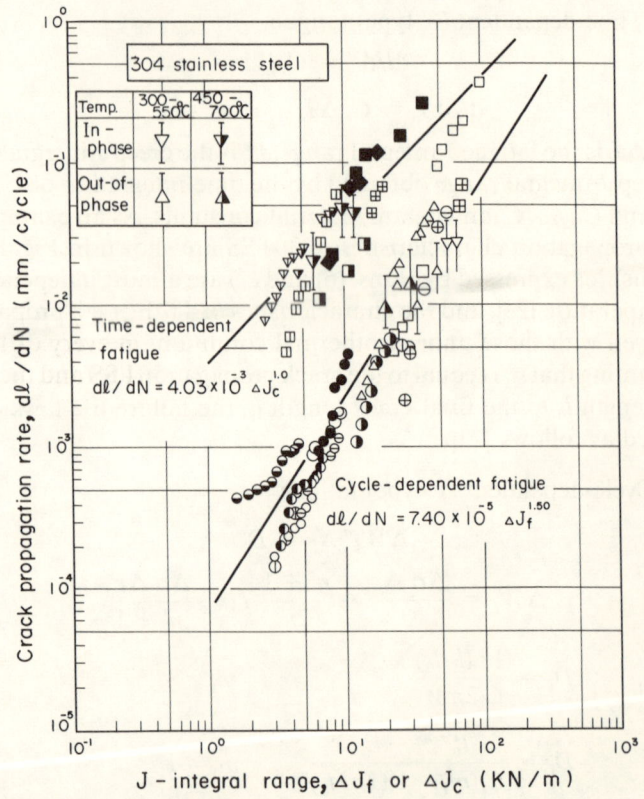

FIG. 10. Macrocrack propagation characteristics of 304 stainless steel for TMF as well as for isothermal fatigue at 550 °C, 600 °C, 650 °C and 700 °C.

where $\sigma_{max}$ and $\Delta\sigma$ are the maximum tensile stress and stress range; $\Delta\varepsilon$, $\Delta\varepsilon_e$, $\Delta\varepsilon_p$ and $\Delta\varepsilon_c$ are the total, elastic, plastic and creep strain ranges; $\dot{\varepsilon}_t$ is the tension-going strain rate; $\beta$ is the stress wave form exponent, $\sigma = \sigma_{max}(t/t_0)^\beta$; $t_0$ is the tension-going time, $t_0 = \Delta\varepsilon/\sqrt{2\dot{\varepsilon}_t}$; $n'$ is the inverse of the work hardening exponent, $\Delta\varepsilon_p \propto \Delta\sigma^{n'}$; $n$ is the creep exponent in Eqn. (5); $f(n) = 3\cdot85\sqrt{n(1 - 1/n)} + \pi/n$; $M_J$ is the boundary correction factor divided by the flaw shape correction factor for $K_I$ (i.e. $M_J = M_k^2/Q$)

Also, $\Delta\tilde{W}_f$ and $\Delta\tilde{W}_c$ are called strain energy parameters. The laws expressed by Eqns. (8) and (9) are similar to the Manson–Coffin relationship and the strain range partitioning relationships. In particular, there should be close correspondence between the derived laws and the existing ones when $\Delta\varepsilon_p$ or $\Delta\varepsilon_c$ is a large fraction of $\Delta\varepsilon$ (i.e. for ductile materials). However, Eqns. (8) and (9) imply that the low cycle fatigue failure life at high temperatures is controlled not only by the plastic or creep strain range, but also by the elastic strain range, the stress range and/or the maximum tensile stress. In fact, the strain energy parameter approach demonstrates its applicability for a superalloy such as Mar M247, in which the contribution to $N_f$ of the elastic strain and the stress is not negligible, while the existing $\Delta\varepsilon_{in}$-based approaches are not available for the life prediction of a superalloy [23]. Furthermore, it should be noted that the controlling variables in Eqns. (8) and (9) are only dependent on the tension-going deformation in every cycle.

Figure 11 shows the relationship between $N_f$ for 304 SS and the strain

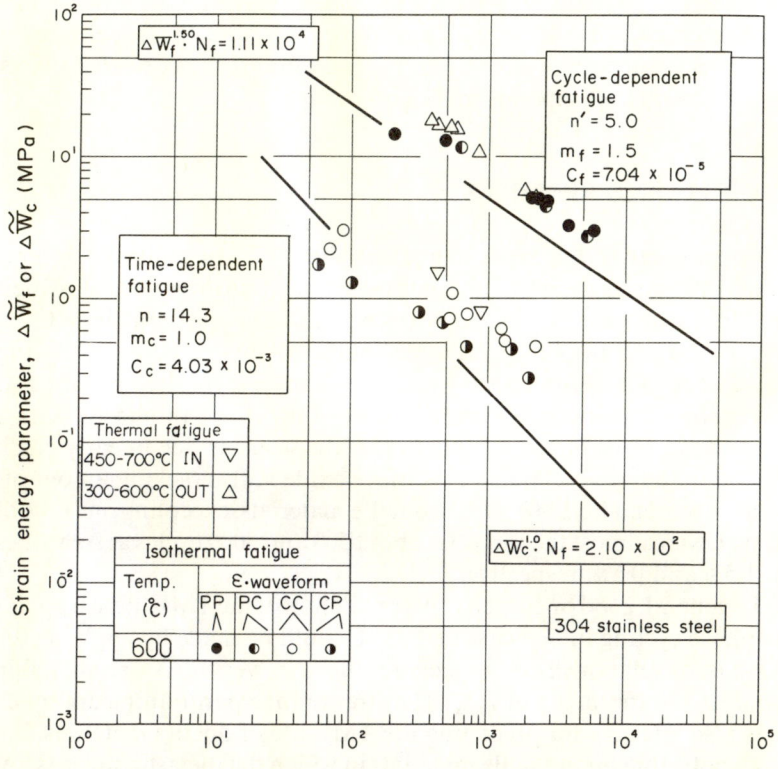

FIG. 11. Relationships between the failure life and strain energy parameters for the thermal–mechanical and isothermal fatigue of 304 stainless steel.

energy parameters in TMF as well as isothermal fatigue with four types of strain waveforms (i.e. fast–fast or PP-type, fast–slow or PC-type, slow–slow or CC-type, slow–fast or CP-type). The relationship is classified into cycle-dependent (P-type: PP- and PC-types, and out-of-phase) and time-dependent (C-type: CC- and CP-types, and in-phase) categories like the crack propagation characteristics. A transgranular fracture with striations occurs in the P-type category, while an intergranular fracture due to creep occurs in the C-type category. Furthermore, Fig. 11 demonstrates the correspondence between the relationship for TMF and that for isothermal fatigue. The solid lines in Fig. 11 represent the strain energy parameter versus $N_f$ relationships predicted from the crack propagation characteristics shown in Fig. 10. In this case, $l_0 = 0.1$ mm as the initial crack length at $N = 0$ and $l_f = 1.5$ mm as the final crack length at $N = N_f$, and $M_J = 0.51$, which is the value for a semi-circular surface crack. The predicted lines agree well with the experimental results. Similar results have been obtained for other materials such as Cr–Mo steels and superalloys [23].

## CONCLUDING REMARKS

To summarize the experimental results for TMF failure in several kinds of steels and alloys, the $\Delta\varepsilon$ versus $N_f$ relationships were classified into four categories, including two types (called Type I and Type O) in which $N_f$ under in-phase and out-of-phase strain–temperature cycling conditions was shorter in a low strain range. Type I is ascribed to intergranular fracture due to the effect of creep for in-phase cycling, while Type O, observed mainly in Cr–Mo steels, is due to early crack initiation that is accelerated by surface oxide scale cracking for out-of-phase cycling. The TMF lives of ductile materials, excepting one case of Type I, were almost the same, i.e. about 500 and 5000 cycles at $\Delta\varepsilon$ values of 1.5% and 0.5%, respectively.

In general, good or conservative prediction of the TMF life, except for in-phase cycling in one case of Type I, could be made by applying the $\Delta\varepsilon$-based relationships, especially the $\Delta\varepsilon$ versus $N_f$ relationship obtained isothermally at $T_{max}$. The strain range partitioning approach was also effective for predicting the TMF life of ductile materials, but not applicable for inductile materials in which the inelastic strain is low because of their high strength.

Based on the experimental finding that high temperature, low cycle fatigue failure was brought on by the propagation of a main crack, the strain energy parameters approach can be derived from the fracture mechanics laws of macrocrack propagation. Since the contributions of the elastic strain and the stress, as well as the inelastic strains, to $N_f$ can be demonstrated by this approach, it is applicable for both inductile and ductile materials. The relationship between $N_f$ and the strain energy parameters was also classified into cycle- and time-dependent categories like the crack propagation characteristics, the relationship for TMF agreeing well with that for isothermal fatigue. Furthermore, this relationship can be predicted from the crack propagation characteristics.

## REFERENCES

[1]   D. A. Spera, *ASTM STP 612*, 3 (1976).
[2]   E. Glenny, J. E. Northwood, S. W. K. Shaw and T. A. Tailor, *J. Inst. Met.*, **87**, 449 (1958–59).
[3]   For example, P. T. Bizon and D. A. Spera, *ASTM STP 612*, 106 (1976).
[4]   L. F. Coffin, Jr., and R. P. Wisley, *Trans. ASME*, **76**, 923 (1954).
[5]   K. Kuwabara, A. Nitta and T. Kitamura, *Proc. ASME Int. Conf. on Advances in Life Prediction Method*, D. A. Woodford and J. R. Whitehead, Eds., ASME, New York, 131 (1983).
[6]   A. Nitta, K. Kuwabara and T. Kitamura, *Proc. 1983 Tokyo Int. Gas Turbine Congr.*, Gas Turbine Society of Japan, Tokyo, 765 (1983).
[7]   S. Taira, M. Fujino and R. Ohtani, *Fatigue Eng. Mater. Struct.*, **1**, 495 (1979).
[8]   M. Fujino, A Basic Study on Thermal Fatigue of Steels, *Doctoral Thesis*, Kyoto University, (1978).
[9]   K. Kuwabara and A. Nitta, CRIEPI Report E277005, Central Research Institute of Electric Power Industry, Tokyo, Japan (1977).
[10]  K. Kuwabara and A. Nitta, *Proc. 1976 ASME–MPC Symp. on Creep–Fatigue Interaction*, R. M. Curran, Ed., ASME, New York, 161 (1976).
[11]  K. Kuwabara and A. Nitta, *Fatigue Eng. Mater. Struct.*, **2**, 293 (1979).
[12]  K. Kuwabara, A. Nitta and T. Kitamura, *J. Soc. Mater. Sci. Jpn.*, **32**, 1167 (1983).
[13]  H. Teranishi and A. J. McEvily, *Metall. Trans.*, **10A**, 1806 (1979).
[14]  K. D. Challenger, A. K. Miller and C. R. Brinkman, *J. Eng. Mater. Technol., Trans. ASME*, **103**, 7 (1981).
[15]  K. Kuwabara, A. Nitta and T. Kitamura, *J. Soc. Mater. Sci. Jpn.*, **32**, 657 (1983).
[16]  S. S. Manson, *Exp. Mech.*, **5**, 193 (1965).
[17]  S. S. Manson, *ASTM STP 520*, 744 (1973).
[18]  S. S. Manson, G. R. Halford and M. H. Hirschberg, *ASME Symp. on Design for Elevated Temperature Environments*, S. Y. Zamrik, Ed., ASME, San Francisco, 12 (1971).

[19] G. R. Halford and J. F. Saltsman, *Proc. ASME Int. Conf. on Advances in Life Prediction Methods*, D. A. Woodford and J. R. Whitehead, Eds., ASME, New York, 17 (1983).
[20] G. R. Halford and S. S. Manson, *ASTM STP 612*, 239 (1976).
[21] A. Nitta, K. Kuwabara and T. Kitamura, *CRIEPI Report E282015*, Central Research Institute of Electric Power Industry, Tokyo, Japan (1983).
[22] S. S. Manson, G. R. Halford and A. C. Nachtigall, in *Advances in Design for Elevated Temperature Environment*, S. Y. Zamrik and R. I. Jetter, Eds., ASME, 17 (1975).
[23] R. Ohtani, T. Kitamura, A. Nitta and K. Kuwabara, Presented at *Int. Symp. Low Cycle Fatigue — Directions for the Future, Lake George, New York, 1985, ASTM STP* (to be published).
[24] A. Nitta, T. Ogata and K. Kuwabara, *Proc. 24th Symp. on Strength of Materials at High Temperatures*, The Society of Materials Science, Japan, 127 (1986).

# Modifications to the Creep–Fatigue Life Criterion for High Temperature Nuclear Components

MASAKI KITAGAWA, ISAMU NONAKA, HIROSHI HATTORI and
AKIRA OHTOMO
*Ishikawajima-Harima Heavy Industries Co. Ltd., 3-1-15 Toyosu Kotoku,
Tokyo 135, Japan*

## ABSTRACT

In order to make adequate use of the linear damage summation criterion that is the basis of most design standards, a detailed experimental study was performed on the creep–fatigue behaviour of actual plant materials. Based on the experimental data, three modifications to the criterion are proposed for use in the life prediction of high temperature nuclear components, namely a modification for the loading wave form effect, a modification for the ratcheting strain effect on high temperature gas-cooled reactor materials, and a proposal for the life prediction method of welded joints in fast breeder reactor materials.

The results of a comparison between the experimental and predicted lives indicate that the accuracy of the prediction is reasonable and was significantly improved by these modifications.

## INTRODUCTION

Accurate prediction of the creep–fatigue life is necessary for assessing the structural integrity of high temperature components. The basic approach to life determination is the linear damage summation method, which was originally proposed by Taira, and is widely used after the rule was accepted in the *ASME Boiler and Pressure Vessel Code, Case N-47* [1].

Since then, the linear damage summation rule has been used as the design criterion for creep–fatigue failure. Although the linear summation rule may not be the most accurate criterion for creep–fatigue failure, it is kept in use in the design procedure because of its simplicity.

The authors have been involved in the development of the design and life prediction procedures for fast breeder reactor (FBR) and high temperature gas-cooled reactor (HTGR) components on the basis of the *ASME Boiler and Pressure Vessel Code, Case N-47*. In these procedures, the creep fatigue failure criteria were modified so that a more accurate life prediction was possible. The improved procedure is mainly based on experimental data for the corresponding materials and environments. Typical modifications are related to the effects of the ratchet strain and the wave form for HTGR components, the effect of the compressive hold time on the creep fatigue life of Cr–Mo steel in an oxidizing atmosphere, and the effect of the loading wave form on the creep–fatigue life of SUS 316 austenitic stainless steel at FBR temperatures. A simplified design procedure for the creep-fatigue life of welded joints was also proposed.

In this chapter some of these activities are summarized for the reference of designers of high temperature components.

## EFFECT OF THE LOADING WAVE FORM ON THE CREEP-FATIGUE LIFE OF HTGR COMPONENTS

### Aim

The primary stresses in HTGR components are usually smaller than the thermal stresses, which do not produce an accumulated strain because of their nature. In such cases, the creep–fatigue life is strongly dependent on the loading patterns (the combination of strain rates). Therefore, the life criterion under various loading patterns for HTGR materials in an HTGR environment was studied.

### Experimental Conditions

Two candidate materials for an HTGR structure, namely Inconel 617 and Hastelloy XR, were tested at 1000 °C and 950 °C, respectively. Their chemical compositions are listed in Table 1, the materials being used in helium gas, which is an inert atmosphere. In order to include the effect of the environment in the life prediction, creep–fatigue tests were performed in helium gas and vacuum conditions, as well as in air. The tests were performed under strain control using trapezoidal waves and saw tooth waves of various frequencies.

### Proposed Procedure for the Creep–Fatigue Life Preduction Under Various Loading Patterns [2]

The criterion proposed for the design procedure is a modification to the linear creep–fatigue damage summation rule which is usually used

TABLE 1
Chemical compositions, ASTM grain size and heat treatments of Inconel 617 and Hastelloy XR

| Element | Chemical composition (wt%) | | | Product form |
|---|---|---|---|---|
| | Inconel 617 (Heat A) | (Heat B) | Hastelloy XR | Inconel 617(Heat A) |
| Carbon | 0·08 | 0·09 | 0·07 | Hot-rolled plate, 20 mm thick |
| Chromium | 21·56 | 21·90 | 21·45 | Inconel 617(Heat B) |
| Nickel | Balance | Balance | Balance | Hot-rolled bar of diameter 25.4 mm |
| Iron | 1·46 | 0·44 | 18·36 | Hastelloy XR |
| Molybdenum | 9·96 | 8·96 | 8·96 | Hot-rolled bar of diameter 20 mm |
| Cobalt | 12·20 | 12·37 | 0·012 | |
| Tungsten | — | — | 0·47 | ASTM grain size |
| Manganese | 0·04 | 0·02 | 0·99 | Inconel 617(Heat A) |
| Titanium | 0·55 | 0·47 | 0·02 | G.S. No. 3.0 |
| Aluminium | 1·01 | 1·28 | 0·01 | Inconel 617(Heat B) |
| Boron | — | — | 0·001 | G.S. No. 3.5 |
| Silicon | 0·13 | 0·14 | 0·25 | Hastelloy XR |
| Phosphorus | — | — | 0·003 | G.S. No. 3.5 |
| Sulphur | 0·005 | 0·002 | 0·006 | Solution heat treatment |
| | | | | Inconel 617 |
| | | | | 1177°C for 1 h, air cooled |
| | | | | Hastelloy XR |
| | | | | 1190°C for 1 h, water quenched |

—, not measured.

in the design standards. The difference from the conventional linear rule is in the method for determining the creep damage. Creep damage in the proposed criterion is determined by using the cyclic creep rupture life instead of the static creep rupture life. Furthermore, the constant $\alpha$ was introduced in order to allow for the difference in damage from a tensile creep history and from a compressive creep history.

The criterion can be expressed by the following equations:

$$\Phi_F + \Phi_C' = D \qquad (1)$$

and

$$\Phi_F = n/N_f \qquad (2)$$

$$\Phi_C' = \alpha\Sigma(\Delta t/t_{r,\,\text{cyclic}})$$

where $\Phi_F$ is the fatigue damage as a function of the applied cyclic strain magnitude, $\Phi_C'$ is the time ratio creep damage modified as a function of the type of strain wave form, i.e. slow–slow (SS), slow–fast (SF) and fast–slow (FS) wave forms, $D$ is the total creep–fatigue/environment interaction damage, $n$ is the number of cycles applied at a particular loading condition, $N_f$ is the corresponding fatigue life for the fast–fast (FF) type of continuous cycling, $\alpha$ is a coefficient reflecting the difference between creep damage by tensile creep and that by compressive creep, $\Delta t$ is the increment of time under a given stress level and $t_{r,\,\text{cyclic}}$ is the cyclic creep rupture time in the corresponding atmosphere.

The cyclic creep rupture time was determined from the result of symmetrical continuous cycling tests (FF and SS types of wave forms). Figure 1 shows a comparison between the monotonic and cyclic creep rupture strengths of both alloys at HTGR temperatures in various environments. The cyclic creep rupture lives were 5–10 times longer than the monotonic creep rupture lives and the different environments affected the monotonic and cyclic creep rupture strengths.

A coefficient $\alpha$ reflected the difference between damage by tensile creep and that by compressive creep. The damage caused by tensile creep differed from the damage by compressive creep, and an SF (with strain hold in tension only) type of wave form produced the most significant creep damage. Life reductions under SF, FS (with strain hold in compression only) and SS (with strain hold on both sides) types of wave forms are defined as $(N_f^0 - N_f^T)$, $(N_f^0 - N_f^C)$, and $(N_f^0 - N_f^{TC})$, respectively. $N_f^0, N_f^T, N_f^C$ and $N_f^{TC}$ are the lives under FF, SF, FS and SS wave forms, respectively. The life reduction ratio $\alpha = (N_f^0 - N_f^i)/(N_f^0 - N_f^T)$

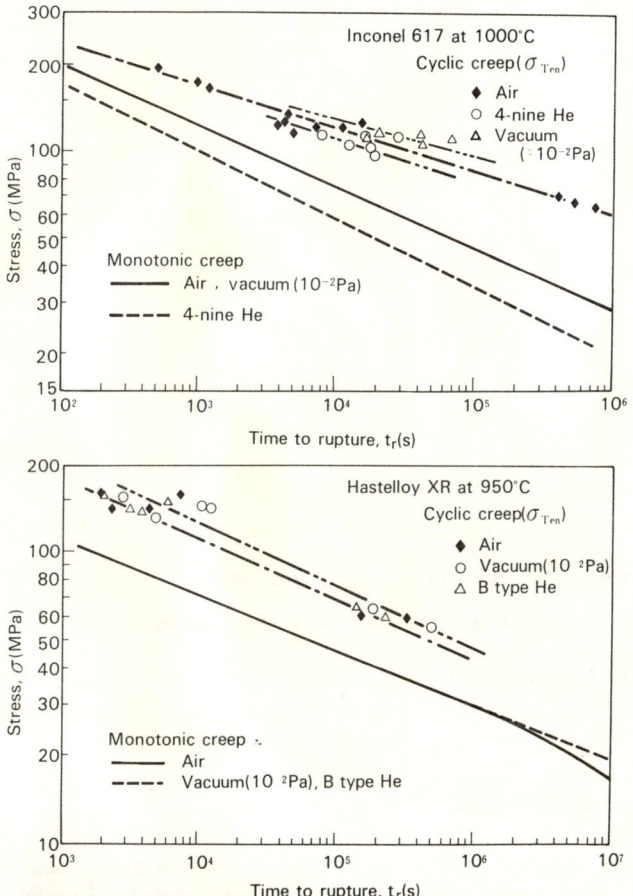

FIG. 1. Monotonic and cyclic creep rupture strengths of Inconel 617 and Hastelloy XR at HTGR temperatures in various environments.

of FS and SS types to the SF type from low cycle fatigue (LCF) tests are shown in Fig. 2.

From these results, the following assumptions could be added to the creep damage.

(1) Damage by tensile creep only

$$\Phi'_C = \Phi^T_C \qquad\qquad \alpha = 1$$

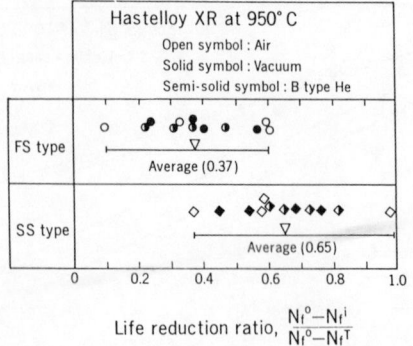

Life reduction ratio, $\dfrac{N_f^o - N_f^i}{N_f^o - N_f^T}$

FIG. 2. Life reduction ratio of FS and SS types relative to the SF type from LCF tests for Hastelloy XR at 950 °C in various environments.

(2) Damage by compressive creep only

$$\Phi_C = \Phi_C^C = \alpha\Phi_C^T = 0.37\Phi_C^T \qquad \alpha = 1/3$$

(3) Damage by both tensile and compressive creep

$$\Phi_C = \Phi_C^{TC} = \alpha\Phi_C^T = 0.65\Phi_C^T = \Phi_C^T - \Phi_C^C \qquad \alpha = 2/3$$

This result suggests that the damage from both tensile creep and compressive creep in one cycle might be recovered from the damage by tensile only creep during the compressive creep regime.

### Results of the Life Estimation

The creep–fatigue damage by the proposed criterion is shown in Fig. 3, the creep–fatigue damage by the conventional linear damage summation rule being shown in Fig. 4. It can be clearly seen that the scatter of damage calculated by the proposed method was smaller than that given by the conventional linear summation rule, the total damage being in the range of 0.3–1.5. Most of the improvement associated with the proposed method was caused by the use of cyclic creep rupture data in the place of static creep rupture data.

## EFFECT OF THE RATCHETING STRAIN ON THE CREEP-FATIGUE LIFE OF HTGR COMPONENTS

### Aim

Some HTGR components are subjected to cyclic thermal stresses as well as the primary stress, resulting in the accumulation of ratcheting

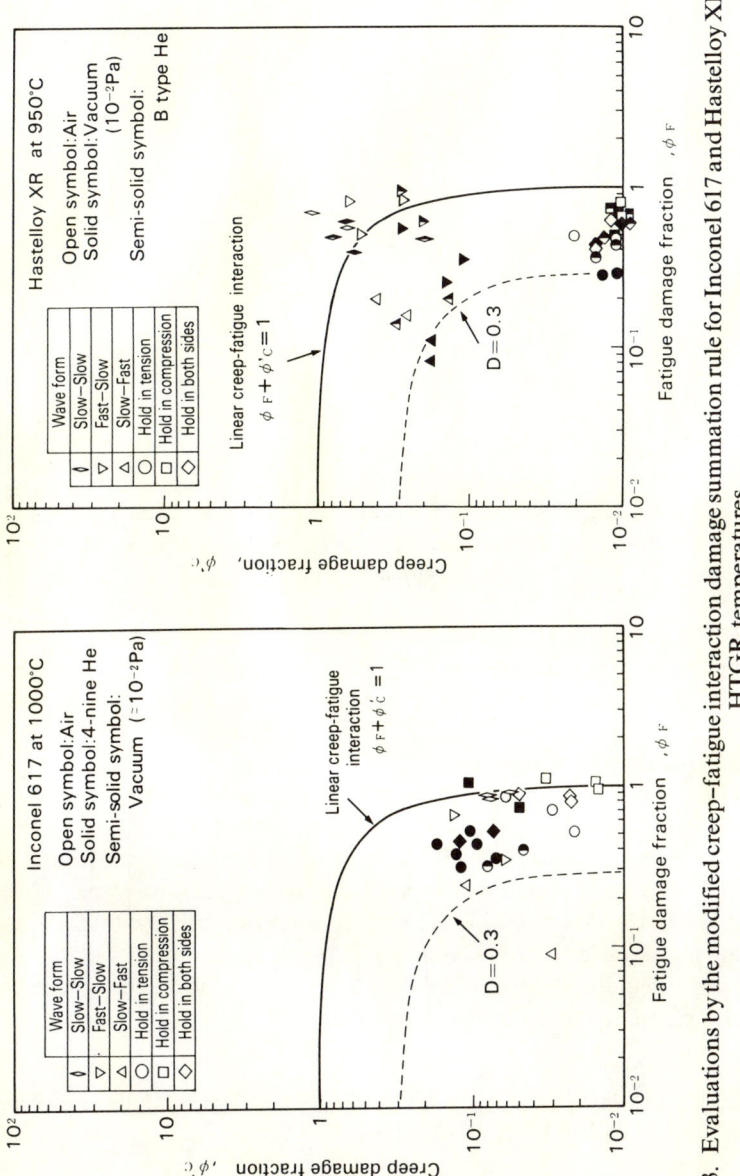

FIG. 3.  Evaluations by the modified creep–fatigue interaction damage summation rule for Inconel 617 and Hastelloy XR at HTGR temperatures.

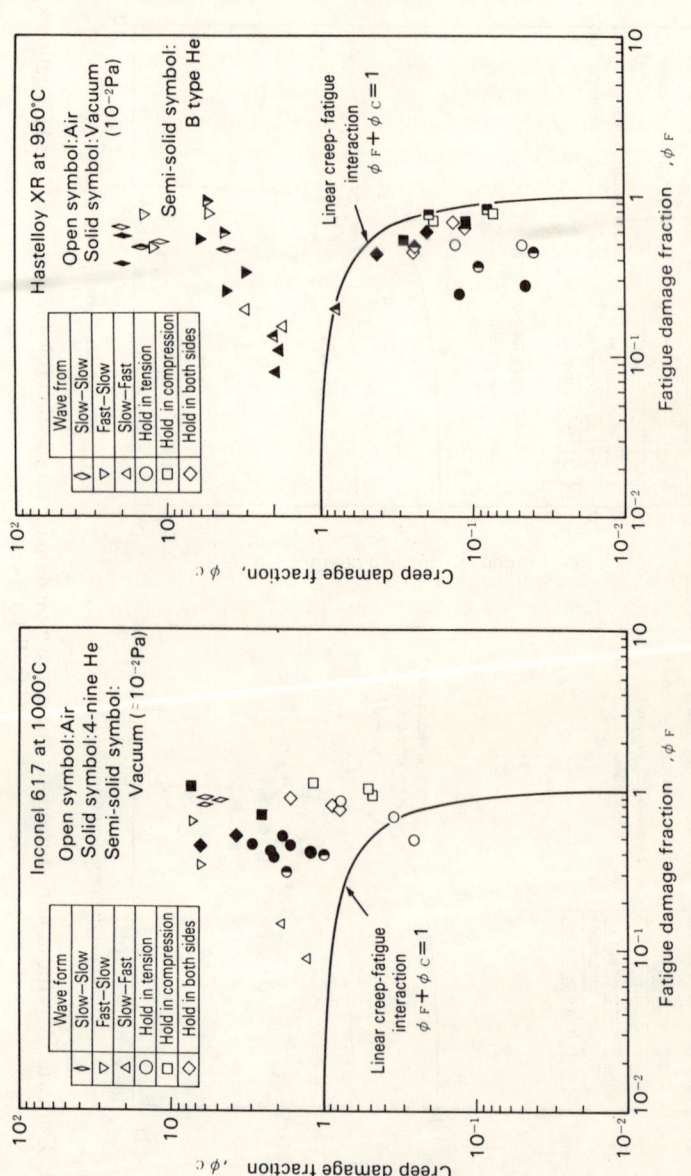

FIG. 4. Evaluations by the conventional creep-fatigue interaction damage summation rule for Inconel 617 and Hastelloy XR at HTGR temperatures.

strain. An example is the stub portion of the center pipes in the helical coil type of intermediate heat exchanger. The primary stress is due to the internal stress, and the cyclic stress is due to the temperature difference between the heat exchanging tube bundles and the center pipe during the start-up and shut-down cycles. Similar situations have been encountered in the metalic skin of an iron-making furnace.

The usual fatigue design standards are based on the assumption that no significant ratcheting strain should occur, and therefore the procedure is not applicable to the life prediction of these cases. In order to modify the life prediction criterion for these cases, the creep fatigue life criterion under ratcheting strain was studied.

## Experimental Conditions

The material selected for the experimental model of an HTGR inter-mediate heat exchanger was the nickel-base superalloy Inconel 617, the maximum operating temperature being 1000 °C. A series of creep-fatigue tests was performed under these conditions, using the loading wave patterns shown in Fig. 5, in order to establish the life criterion under a ratcheting strain. Besides these experiments, another series of tests in an HTGR atmosphere (namely helium gas) was performed to determine the material properties in the HTGR atmosphere. These will

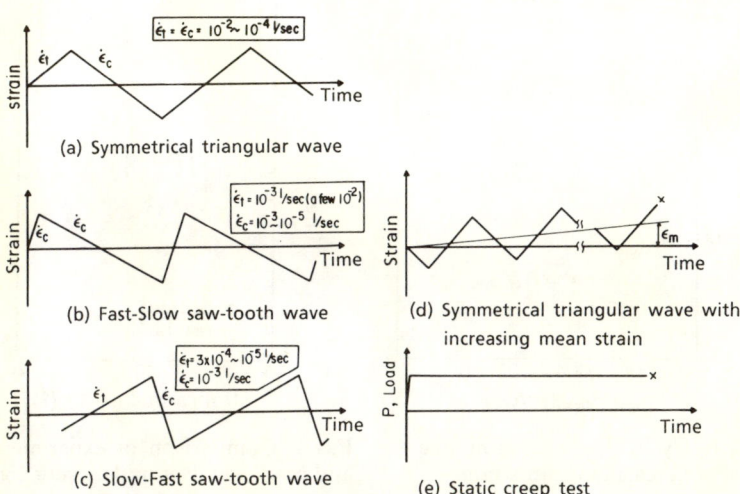

FIG. 5. Loading patterns employed for the laboratory tests.

be used when the criterion is applied to evaluate the life to fracture of the partial model of an intermediate heat exchanger.

## Proposed Life Prediction Procedure [3, 4]

Figure 6 shows the creep–fatigue failure life under a symmetrical triangular wave with increasing mean strain. It can be seen that the creep–fatigue life decreased with increasing mean strain per cycle. The same results are plotted as a relationship between the time to fracture and the unidirectionally accumulated strain in Fig. 7. The decrease of time to failure seems to be proportional to

$$\Phi/\text{cycle} = \Delta t/t_r, \text{cyclic} + \alpha\varepsilon_m$$

where $t_{r,\text{cyclic}}$ is the cyclic rupture life, $\varepsilon_m$ is the unidirectionally accumulated strain per cycle, and $\alpha$ is a material constant.

Two extreme cases of this criterion are unidirectional creep rupture and cyclic creep rupture. After applying these two cases, the constant $\alpha$ could be determined using the static and cyclic creep rupture lives ($t_{r,\text{static}}$

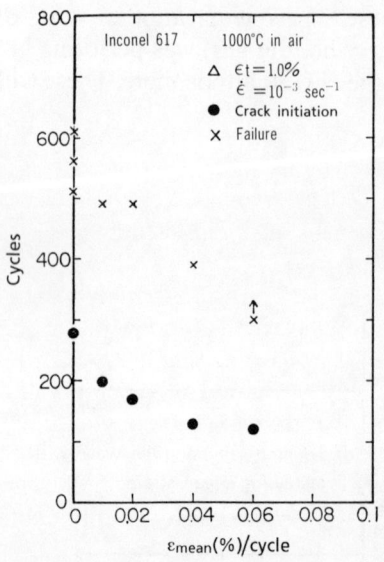

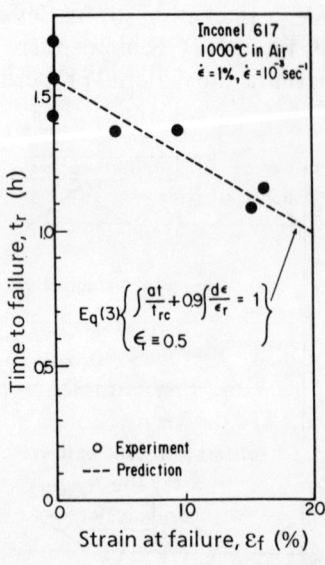

FIG. 6. Cyclic life reduction due to increasing mean strain.

FIG. 7. Comparison of experimental and predicted lives under cyclic loading with increasing mean strain.

and $t_{\text{r, cyclic}}$). Then, the life criterion under accumulated strain can be determined by the following equation:

$$1 = \Delta t / t_{\text{r, cyclic}} + (t_{\text{r, cyclic}} - t_{\text{r, static}}) \cdot \varepsilon_{\text{m}} / t_{\text{r, cyclic}} \cdot \varepsilon_{\text{r}} \tag{3}$$

Each material property needs to be determined under the corresponding atmosphere. In Fig. 7, the prediction of life by this equation is also shown, indicating that the criterion was reasonably accurate.

## Results of the Application to the Lifetime Test of an Intermediate Heat Exchanger

Although the foregoing criterion was developed on the basis of relatively short-term tests, this criterion was applied to the life evaluation of a partial model of an experimental HTGR intermediate heat exchanger. The model was fractured at high temperature (1000 °C) in helium under an accelerated condition which lasted for about 1000 h and the results were compared with the estimation.

The model was a portion of a helical type of heat exchanger, and consisted of eight heat exchanging tubes and a center pipe. In order to simulate the thermal strain due to the temperature difference between the tube bundle and the center pipe, the center pipe was mechanically moved up and down. Each tube was fixed to the shell at one end, the other ends being welded to the center pipe through the stubs on the center pipe. In order to apply various primary and secondary stresses to all of the tubes, various internal pressures were applied to the tubes and the distance between the stub and the shell weld of each tube was varied.

To evaluate the actual stress in the critical position (which was the stub portion in this model), the finite element method with several simplifications and with many material properties was used.

The results of a comparison between the evaluated and experimental results are shown in Table 2. Pipes 1, 2, 4 and 5 were broken and pipe 3 was found to be cracked from the inside surface during a detailed examination after the fracture test. It can be seen that the total damage to the broken pipes was close to pipe 1. This criterion may be applicable to the rather wide variation of loading histories, because the ratios between creep damage and ductility damage in the fracture tests varied from 0·2 to 2, although the estimation agreed well with the experimental results.

TABLE 2

Comparison of the experimental lives with the lives predicted for the stubs on the center pipe of HTGR intermediate heat exchanger model under the accelerated conditions

| TP number | Predicted results | | | | | | Experimental rupture time $(t_f)_{exp}(h)$ |
|---|---|---|---|---|---|---|---|
| | Outer surface | | | Inner surface | | | |
| | $\phi_c$ | $\phi_D$ | Failure time $(t_f)_o(h)$ | $\phi_c$ | $\phi_D$ | Failure time $(t_f)_i(h)$ | |
| 1 and 2 | 0·620 | 0·380 | 793 | 0·652 | 0·348 | 1033 | 695 and 666 |
| 3 | 0·359 | 0·641 | 787 | 0·404 | 0·596 | 1117 | >700 |
| 4 | 0·209 | 0·891 | 421 | 0·245 | 0·755 | 557 | 400 |
| 5 | 0·153 | 0·847 | 341 | 0·184 | 0·816 | 445 | 372 |

## SIMPLIFIED LIFE PREDICTION PROCEDURE FOR A GROUND-OFF WELDED JOINT

### Aim

The welded joints generally include the deformation strength heterogeneity. The weak portion of the heterogeneity often experiences a larger strain than the average due to the strain concentration, which causes premature failure in the welded joints. Although the usual design procedure includes the evaluation methods of strain concentration related to the weld toe, the evaluation methods for material deformation strength heterogeneity are not suggested.

In usual welded joints, where a significant stress concentration exists at the weld toe, the factor controlling the life of the weld may be the level of the stress concentration. However, the extra weld metal was often smoothed out for such a reason as reduction of the stress. An example of such a weld is the tube-to-tube joint. In these cases, the creep–fatigue life is often determined by the degree of material heterogeneity, and therefore the life prediction of the smoothed welded joints was studied.

### Experimental Conditions

The materials tested were 304 stainless steel weldment and normalized and tempered 2¼Cr–1Mo steel weldment with and without a post-weld heat treatment (PWHT) for use with an FBR. The PWHT condition was 720°C for 8·4 h. The shape and dimensions of specimens and the welding processes are shown in Fig. 8. The welding processes involved were shielded metal arc welding (SMAW), electron beam welding (EBW), non-filler tungsten inert gas arc welding (TIG) and submerged arc welding (SAW).

All of the fatigue tests were conducted in air, using a triangular strain wave form with a constant strain rate of $10^{-3}$ s$^{-1}$ and a trapezoidal strain wave form of 10 min and 60 min holding times. The axial gross strain was controlled over a gage length of 25 mm on the specimen, including the base metal and weld metal. The test temperature was the design temperature of an FBR, namely 550°C for 304 stainless steel and 470°C for 2¼Cr–1Mo steel. Heating was accomplished using an induction coil.

### Proposed Method [5, 6]

The process for life prediction is shown in Fig. 9, commencing with the fatigue damage calculation. Let us suppose that the stress applied to

| Specimen type | Dimensions | Lw/GL | Welding process |
|---|---|---|---|
| I | 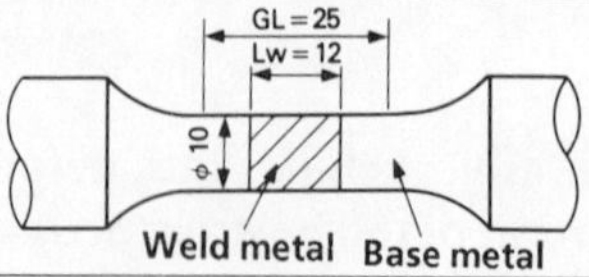 GL = 25, Lw = 12, φ 10, Weld metal  Base metal | 0.5 | SMAW |
| II | 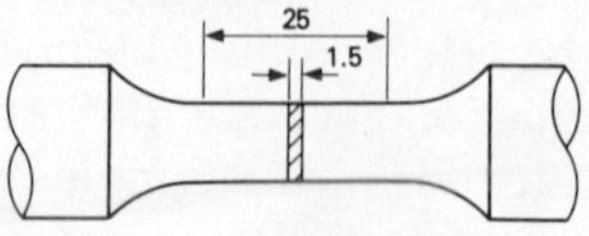 25, 1.5 | 0.06 | EBW |
| III | 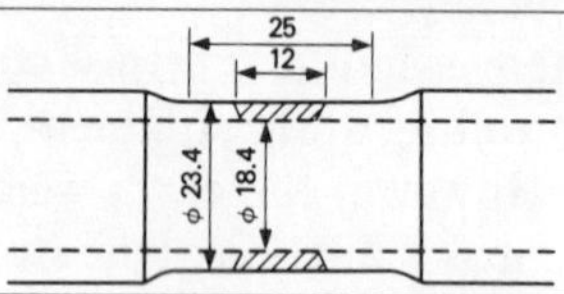 25, 12, φ 23.4, φ 18.4 | 0.5 | TIG |
| IV | 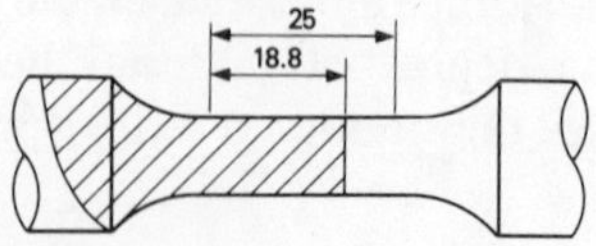 25, 18.8 | 0.75 | |
| V | 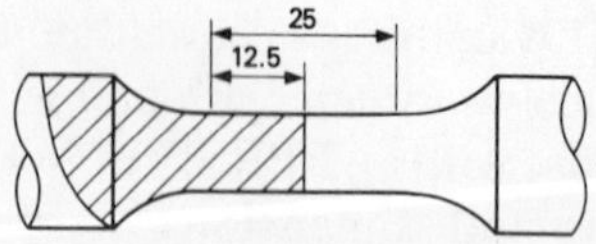 25, 12.5 | 0.5 | SAW |
| VI | 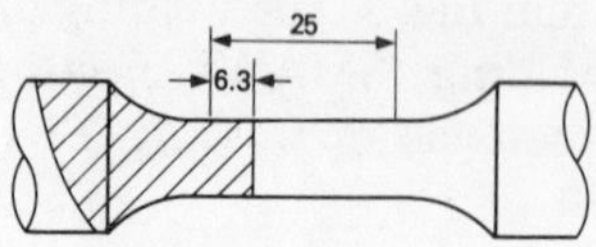 25, 6.3 | 0.25 | |

FIG. 8. Shape and dimensions of the test specimens.

the weldment is given. Using the cyclic stress–strain curve of the all-weld metal and base metal, the average strain in the weld metal region and base metal region were estimated as $\varepsilon_W$ and $\varepsilon_B$. The expected fatigue lives $N_W$ and $N_B$, which correspond to $\varepsilon_W$ and $\varepsilon_B$, were then estimated using the fatigue curves of the all-weld metal and the base metal. When the gross strain across the weldment is given, the stress can easily be estimated by the alternative process shown in Fig. 10. When the fatigue life of the weldment is required, the lesser of the $N_W$ and $N_B$ values just obtained may be selected.

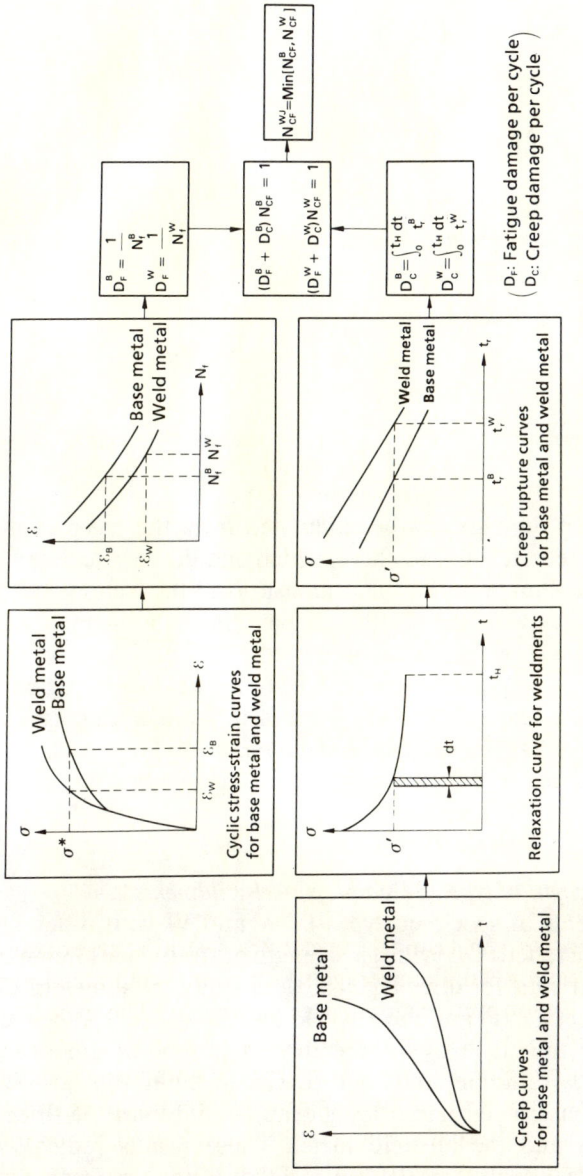

FIG. 9. Creep–fatigue life prediction procedure for weldments.

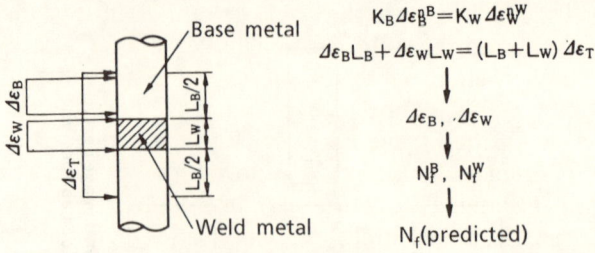

$$K_B \Delta\varepsilon_B{}^B = K_W \Delta\varepsilon_W{}^W$$

$$\Delta\varepsilon_B L_B + \Delta\varepsilon_W L_W = (L_B + L_W)\, \Delta\varepsilon_T$$

$$\downarrow$$

$$\Delta\varepsilon_B,\ \cdot\Delta\varepsilon_W$$

$$\downarrow$$

$$N_f^B,\ N_f^W$$

$$\downarrow$$

$$N_f(\text{predicted})$$

| | | |
|---|---|---|
| $L_B$ | : | Length of base metal region |
| $L_W$ | : | Length of weld metal region |
| $\Delta\varepsilon_B$ | : | Average strain in base metal region |
| $\Delta\varepsilon_W$ | : | Average strain in weld metal region |
| $\Delta\varepsilon_T$ | : | Gross strain controlled |

FIG. 10. Alternative method for the fatigue life prediction of weldments (method II).

The creep damage can be calculated from the creep rupture data, using the creep relaxation curve, which should be calculated from the creep curve equation and the dimensions of the weldment. In the life prediction shown later in this paper, the experimentally obtained relaxation curves for the actual weldment specimens were used to calculate the creep damage $D_C^W$ and $D_C^B$ of the weld metal region and base metal region, respectively. The sum of the fatigue damage and creep damage $(D_f + D_C)$ for each weld metal region and base metal region was then calculated to determine the life of the weldment.

## Results of the Life Prediction

### Fatigue strength of type 304 stainless steel weldments

Weldments of specimen types IV, V and VI with different $L_W/GL$ values were tested (GL is the gage length of strain measurement). Cyclic life comparisons for three types of weldments are shown in Fig. 11 for the base metal data and the all-weld metal data. As can be seen in this figure, the fatigue lives for the three types of weldments were not significantly different, and were similar to the life for all-weld metal. Also, the base metal had a longer fatigue life than the three types of weldments and the all-weld metal. These results indicate that 304 stainless steel weldments deform as if they were homogeneous, because there was little difference in the cyclic deformation resistance between the base metal and the weld metal. The fact that all specimens were fractured in the weld metal region was expected since the fatigue life of the all-weld metal was shorter than that of the base metal.

The results of the fatigue life prediction are shown in Fig. 12. The prediction started under the assumption that the stress was given. It can be seen that the fatigue life of the weldments could be predicted within a factor of 2 on life, the accuracy of the prediction being independent of the $L_w$/GL ratio. Similar results were obtained by the alternative method of Fig. 10, and the predicted results by a finite element method (FEM) analysis are also shown in Fig. 12.

Despite the simplicity of the proposed method compared with FEM analysis, the accuracy of prediction by the proposed method is as good as that of FEM analysis.

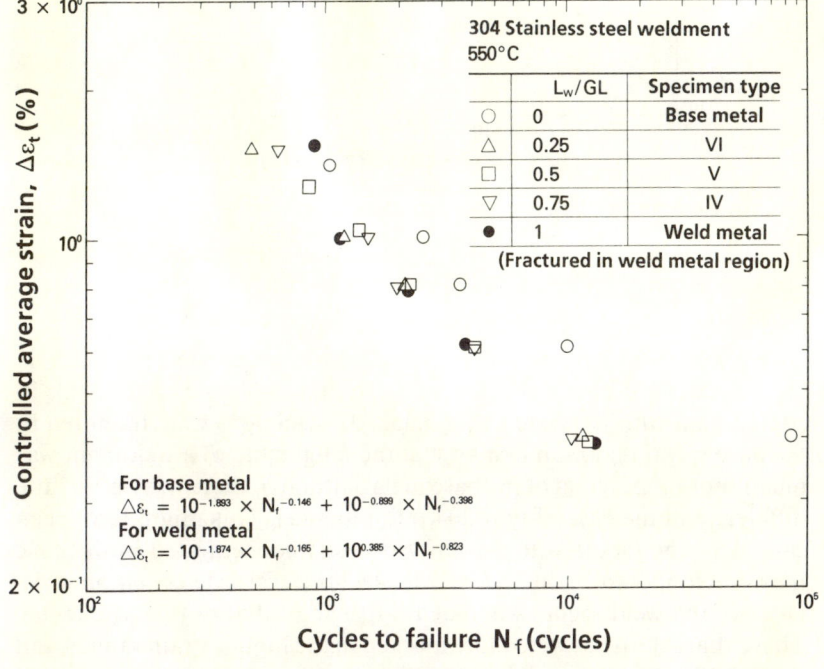

FIG. 11. Fatigue life comparisons for three kinds of type 304 stainless steel weldments.

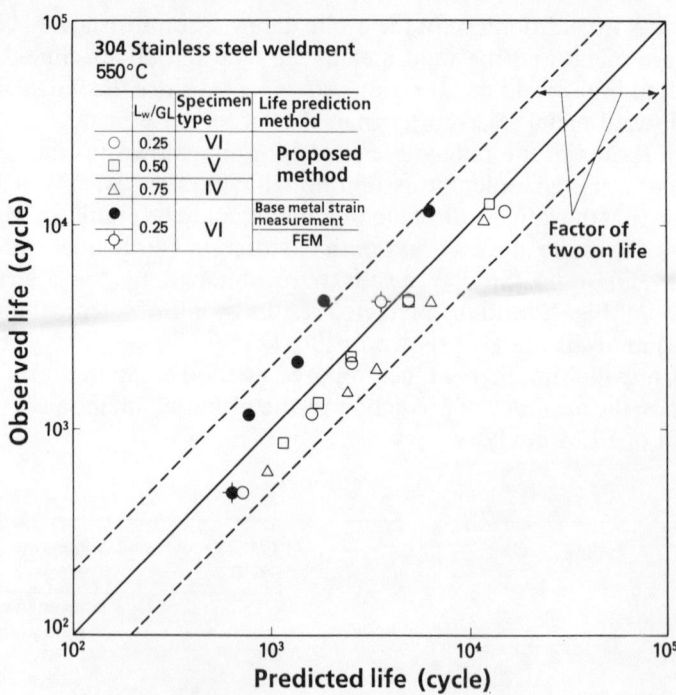

FIG. 12. Comparison between experimental and predicted fatigue lives for type 304 stainless steel weldments.

## Fatigue strength of 2¼Cr–1Mo steel weldments

Weldments of specimen types I, II, III and V were tested. An example of the fatigue life of a weldment (type III) is shown in Fig. 13, in comparison with the base metal data. All specimens were fractured in the base metal region. It can be that the fatigue life of a weldment was much shorter than that of the base metal, although there was only a little difference in the case of type 304 stainless steel. This might have been caused by the fact that the strain was easily concentrated in the base metal region, because in 2¼Cr–1Mo steel the cyclic deformation resistance of the weld metal was much larger than that of the base metal. These characteristics were more marked in the higher strain ranges, and the difference in fatigue life between the weldments and the base metal was greater in the higher strain ranges. Also, it can be seen in Fig. 13 that the fatigue lives show a crossover near $10^3$ cycles. This crossover can be qualitatively estimated from the proposed prediction method. As for the type I and type III specimens, the life prediction was overconservative for a weldment without PWHT, while the life prediction for a weldment with PWHT could be performed within a factor of 2 on life, although it was still unconservative. For the type II specimen, the life prediction was made within a factor of 2 on life, and was independent of the PWHT.

The overconservativeness of the life prediction for the type I and type III specimens without PWHT may have been caused by plastic restraint due to the great difference in cyclic deformation resistance between the base metal and the weld metal. Consequently, an FEM analysis and examination of the fracture location were performed to evaluate this plastic restraint.

The results of the FEM analysis show that the axial strain in the base metal region for the type I specimen was higher than the controlled gross strain in the section away from the weld fusion line. This was evident especially for the type I specimen without PWHT.

The fracture location was in the base metal region away from the weld fusion line for the weldment without PWHT, while it was in the region near the weld fusion line for the weldment with PWHT. This characteristic could be predicted by the results of the FEM analysis, if the fracture location was supposed to be at the maximum axial strain spot.

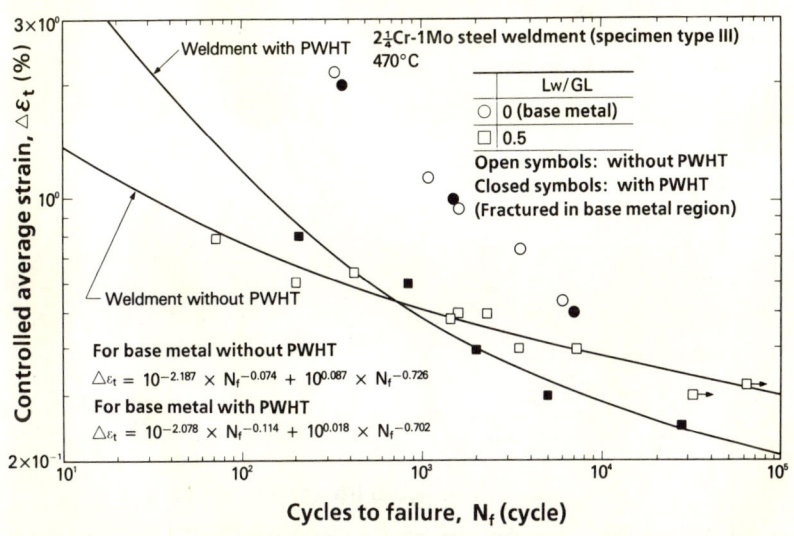

FIG. 13. Fatigue lives for three kinds of 2¼Cr–Mo steel weldments.

FEM analysis and an examination of the fracture location made it clear that distinct plastic restraint occurred in the type I and type III specimens without PWHT.

The reason for the proposed method being overconservative is related to this plastic constraint in the weldment, because the deformation resistance of the base metal was increased. The method presently employed for calculating strain in the base metal started from the experimental stress amplitude. On the other hand, the cyclic stress–strain curves were those obtained for smooth specimens. Consequently, the strain estimated from the experimental stress amplitude and the usual cyclic stress–strain curve must have been larger than the actual strain developed. In other words, the estimated gross strain would have been larger than the controlled gross strain.

Figure 14 shows the results predicted by the alternative method (method II) shown in Fig. 10. It can be seen that the predicted results are

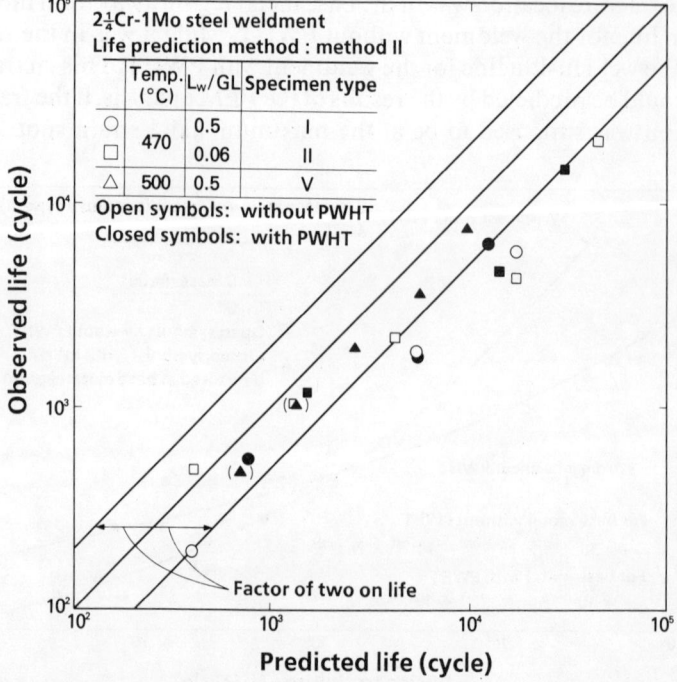

FIG. 14. Comparison between experimental and predicted fatigue lives for 2¼Cr–Mo steel weldments by the alternative method (method II) of Fig. 10.

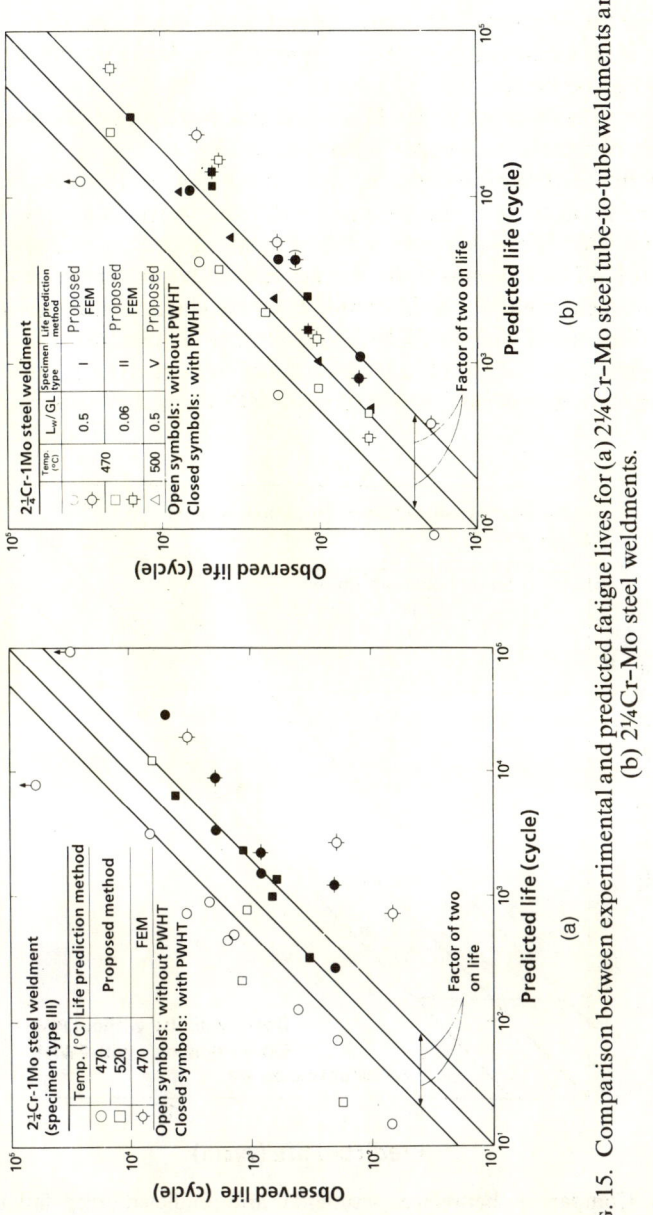

FIG. 15.  Comparison between experimental and predicted fatigue lives for (a) 2¼Cr–Mo steel tube-to-tube weldments and (b) 2¼Cr–Mo steel weldments.

in good agreement with the experimental results. In this method, the gross strain is given first. Consequently, the error caused by plastic constraint is small, enabling the accuracy of prediction by the alternative method to be better.

The results indicate that, when the strain was known, the prediction method of Fig. 10 was better than that of Fig. 9.

Finally, the life prediction based on the FEM analysis (fracture was assumed to occur at the maximum axial strain spot) is shown in Fig. 15. The predicted life was overstated for the type I and type III specimens without PWHT. It seems that this was caused by the strain translation from the weld metal to the base metal, i.e. the elastic follow-up. The life prediction based on the FEM analysis may be used as a simplified design procedure. In such a case we have to bear in mind that it will overestimate the value for a weldment with distinct plastic restraint.

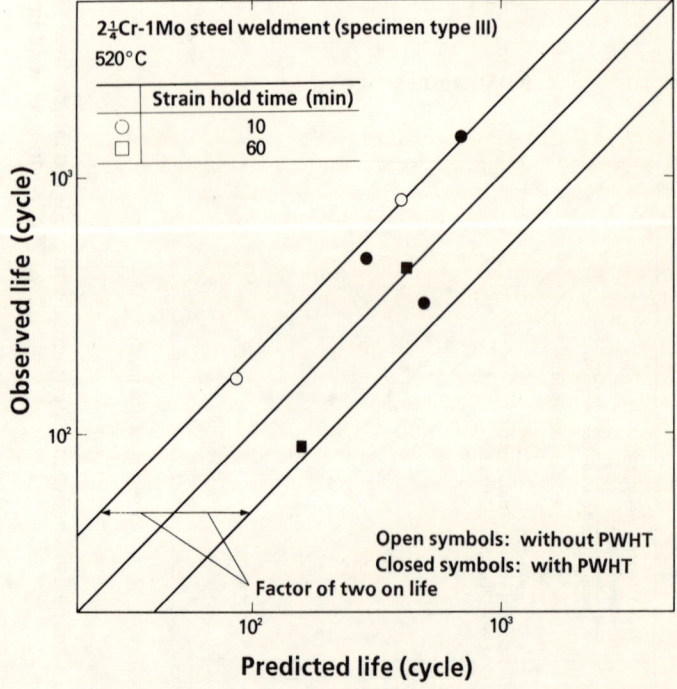

FIG. 16. Comparison between experimental and predicted creep–fatigue lives for 2¼Cr–Mo steel weldments.

*Creep–fatigue strength of 2 ¼Cr–1Mo steel weldments*

The creep–fatigue tests were performed at 520 °C with strain holds of 10 min and 60 min. Only a few data for the type III specimen were obtained, and all the specimens failed in the base metal region.

The predicted results are compared with the experimental results in Fig. 16, and it can be seen that the accuracy of the prediction is within a factor of 2 on life.

## CLOSING REMARKS

Three modifications to the linear damage summation rule for creep–fatigue life prediction are proposed for particular application to nuclear plants in order to improve the accuracy of prediction. The results show that the proposed methods seem to predict the life within factor of approximately 2, which would otherwise not be possible.

## REFERENCES

[1] *ASME Boiler and Pressure Vessel Code, Case N-47,* ASME.
[2] H. Hattori, M. Kitagawa and A. Ohtomo, An evaluation of creep–fatigue/ environment behaviors of Inconel 617 and Hastelloy XR for HTGR Application, *Proc. Int. Conf. on Creep, April 14-18, 1986,* JSME, Tokyo, 117–122.
[3] M. Kitagawa, K. Tamura and A. Ohtomo, A fracture criterion of Inconel 617 under cyclic straining at high temperatures, *J. Soc. Mater. Sci. Jpn.,* **32,** (357) 662–666, (1983).
[4] M. Kitagawa, H. Hattori, A. Ohtomo, T. Teramae and J. Hamanaka, Lifetime test of a partial model of a high temperature gas-cooled reactor helium–helium heat exchanger, *Nucl. Technol.,* **66,** 675–684 (1984).
[5] I. Nonaka, M. Kitagawa and A. Ohtomo, Evaluation of low-cycle fatigue strength for weldments at elevated temperatures — proposal of evaluation method and its application to FBR components, *Proc. of 5th Int. Conf. on Pressure Vessel Technology, San Francisco, September 9-14, 1984,* Vol. II, ASTM, Philadelphia, 1088-1108.
[6] I. Nonaka, *Ph. D. Thesis,* Tohoku University, September 1986.

# Creep and Fatigue Life Prediction Based on the Non-destructive Assessment of Material Degradation for Steam Turbine Rotors

KAZUSHIGE KIMURA, KAZUNARI FUJIYAMA AND
MASAMITSU MURAMATSU
*Heavy Apparatus Engineering Laboratory, Toshiba Corporation,*
*1-9 Suehiro-cho, Tsurumi-ku, Yokohama-city, Japan*

## ABSTRACT

An investigation of steam turbine rotors retired after long-term service at elevated temperatures clarified that one of the most typical characteristics of material degradation in Cr–Mo–V rotor steel was softening, and that the mechanical properties necessary for life prediction such as creep and fatigue were affected by material degradation (softening). Quantitative relationships were established between the material hardness and these degraded mechanical properties. Creep and fatigue life predictions for steam turbine rotors could be more accurately made through the non-destructive measurement of hardness and an evaluation of the degraded creep and fatigue properties.

## INTRODUCTION

Fossil-fuel power plants that have been operated for more than their design lives are increasing in number in Japan. The steam turbine components in these plants are considered to have suffered from material degradation as well as creep and fatigue damage during long-term service at high temperatures. Damage accumulation would be accelerated under the influence of this material degradation and the current operating pattern which has been changed from base load to peak demand operation. In order to maintain a stable power supply, a more accurate life prediction method should be established to enable scheduled maintenance and life extension strategies to be applied to old steam turbine components.

This chapter describes an investigation of the material degradation in steam turbine rotors forged from Cr–Mo–V steel after long-term service at high temperatures. Correlations are made between this material degradation and the mechanical properties particularly related to creep and fatigue, and a new procedure is proposed to predict creep and fatigue damage utilizing non-destructive information about material degradation.

## INVESTIGATION OF MATERIAL DEGRADATION

### Materials

In order to identify material degradation phenomena in Cr–Mo–V rotor forgings during long-term service at high temperatures, a retired high pressure section rotor and high–intermediate pressure section rotors were investigated. Virgin Cr–Mo–V steel and laboratory-aged rotor samples were also tested. The chemical composition, the mechanical properties at room temperature of the original productions and the service history are summarized in Tables 1, 2 and 3, respectively.

TABLE 1
Chemical composition

| Material | C | Si | Mn | P | S | Ni | Cr | Mo | V |
|---|---|---|---|---|---|---|---|---|---|
| Virgin rotor A | 0·27 | 0·27 | 0·66 | 0·009 | 0·007 | 0·34 | 1·10 | 1·23 | 0·25 |
| Retired rotor B | 0·35 | 0·34 | 0·78 | 0·019 | 0·026 | 0·44 | 1·03 | 1·08 | 0·22 |
| Retired rotor C | 0·33 | 0·34 | 0·73 | 0·018 | 0·024 | 0·49 | 0·85 | 1·13 | 0·28 |
| Retired rotor D | 0·32 | 0·33 | 0·71 | 0·015 | 0·021 | 0·49 | 0·99 | 1·16 | 0·26 |

### Test Procedure

*Specimen sampling*

Test samples for creep, creep rupture and low cycle fatigue tests were taken from various locations in the rotors as shown in the example of Fig. 1. Specimens were taken parallel to the rotor axis from the rotor

TABLE 2
Room temperature mechanical properties of original production

| Material | Tensile strength $\sigma_B$ (MPa) | Yield strength $\sigma_{YS}$ (MPa) | Uniform elongation $\varepsilon_u$ (%) | Reduction of area $\phi$ (%) | Heat treatment |
|---|---|---|---|---|---|
| Virgin rotor A | 810 | ·657 | 20·8 | 55·9 | Q: 970°C × 19 h FAN<br>T: 670°C × 52 h FC<br>SR: 640°C × 23 h FC |
| Retired rotor B | 801 | 577 | 21·0 | 51·9 | Q: 955°C × 20 h FAN<br>T: 685°C × 50 h FC<br>SR: 600°C × 16 h FC |
| Retired rotor C | 805 | 629 | 17·9 | 42·9 | Q: 950°C × 21 h FOG<br>T: 670°C × 53 h FC<br>SR: 650°C × 50 h FC |
| Retired rotor D | 793 | 622 | 19·6 | 49·8 | Q: 955°C × 20 h FOG<br>T: 675°C × 60 h FC<br>SR: 650°C × 56 h FC |

$\sigma_{YS}$, 0·02% offset yield strength.
FAN, air cooled; FC, furnace cooled; FOG, mist cooled.

TABLE 3
Aging or service history

| Material | Temperature (°C) | Time (h) |
|----------|------------------|----------|
| Virgin rotor A | 620 | 4 000 6 400 |
| Retired rotor B | Max. 520 | 140 000 |
| Retired rotor C | Max. 536 | 134 000 |
| Retired rotor D | Max. 516 | 130 000 |

surface and also parallel to the circumference from the bore. All the tests were conducted on round bar specimens. Figure 2 shows the dimensions of each specimen.

*Hardness test*

A micro-Vickers hardness tester was used for measuring the material hardness. The hardness was measured at room temperature at various locations on the retired rotors before the mechanical tests. For the laboratory-aged rotor samples, the hardness was measured during aging after specific time intervals.

*Creep and creep rupture test*

Lever-type creep test machines were used for the creep and creep rupture tests, which were conducted at elevated temperatures of 450 °C to 600 °C in air using electric furnaces. Creep elongation was measured with a linear voltage differential transducer extensometer.

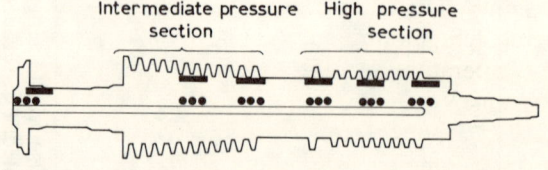

Intermediate pressure section    High pressure section

• Tangential to rotor radial direction
▬ Parallel to rotor axis

FIG. 1. Specimen sampling (example of retired rotor B).

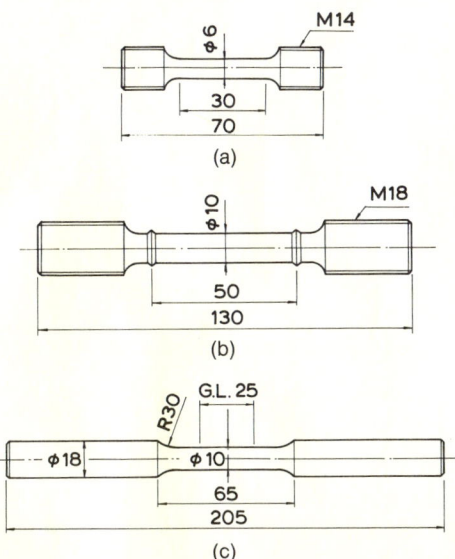

FIG. 2. Specimen dimensions (in millimeters): (a) rupture; (b) creep rate; (c) low cycle fatigue.

## Low cycle fatigue test

Low cycle fatigue tests were conducted by using electro-hydraulic computer-controlled fatigue testing systems. The specimens were heated to 566 °C by induction heating devices, and the axial total strain range of a 25 mm gauge length was controlled. Cyclic stress–strain relationships were obtained by the incremental step method [1].

## Test Results

### Hardness

Figure 3 shows the relationship between the hardness and aging time at each aging temperature for laboratory-aged rotor samples. The higher the temperature and the longer the aging time, the lower the hardness became.

The Vickers hardness levels $H_v$ in the retired rotors were different from each other, ranging from about $H_v = 170$ to $H_v = 260$. Figure 4 shows an example of the radial distribution of hardness for retired rotor B, the hardness being almost constant in the radial direction of the rotor and the hardness of the high temperature portion being lower than that of the low temperature portion.

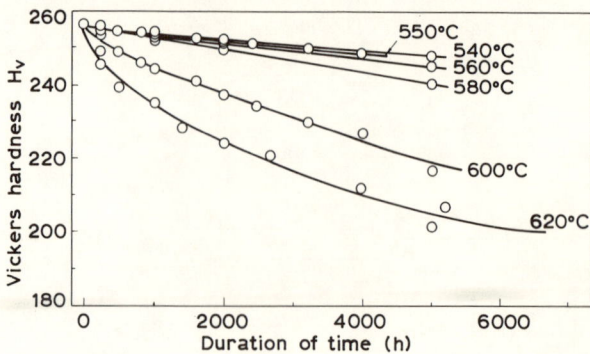

FIG. 3. Time dependence of hardness for a laboratory-aged Cr–Mo–V rotor forging.

*Creep rupture and creep properties*

Creep rupture data for all four rotors are shown in Fig. 5. The Larson–Miller parameter $P$ is expressed as

$$P = (T + 273)(20 + \log t_r) \tag{1}$$

where $T$ is the test temperature (°C) and $t_r$ is the creep rupture time (h). Creep rupture data for the virgin rotor and low temperature portion (LT) of the retired rotors were almost the same as those in the NRIM database [2]. On the other hand, creep rupture data for the aged rotor and high temperature portion (HT) of the retired rotors were far lower than those in this database.

Figure 6 shows the relationship between the minimum creep rate and the applied stress. The aged rotor and high temperature portion of the retired rotors had a faster minimum creep rate than the virgin rotor and low temperature portion of the retired rotors. The latter agree with the data from other authors [3, 4].

*Low cycle fatigue properties*

Figure 7 shows the low cycle fatigue test results. The virgin rotor and low-temperature portion of the retired rotors had almost the same low cycle fatigue strength and the results agree with the data from other authors [5, 6]. The aged rotor and the high temperature portion of the retired rotors were a little lower in low cycle fatigue life than the virgin rotor and low temperature portion of the retired rotors.

Figure 8 shows the cyclic stress–strain relationship obtained for all

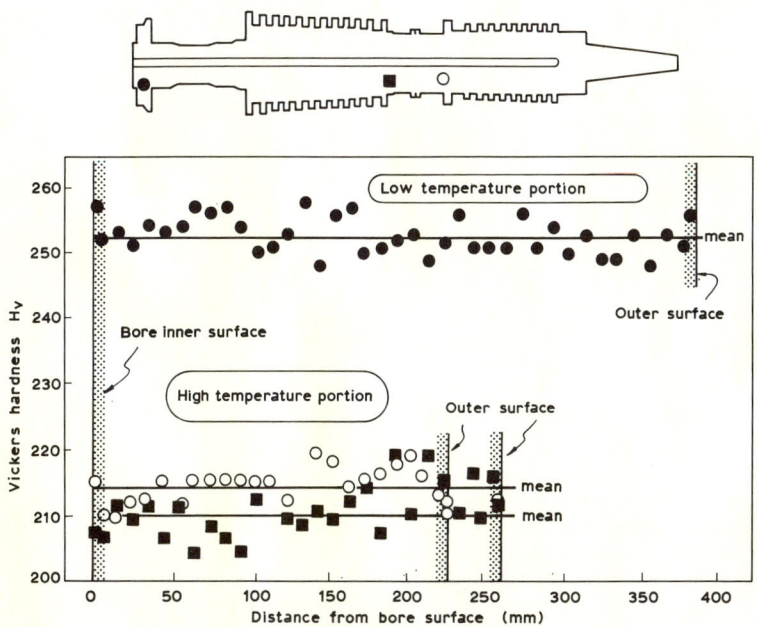

FIG. 4. Radial distribution of Vickers hardness (example of retired rotor B).

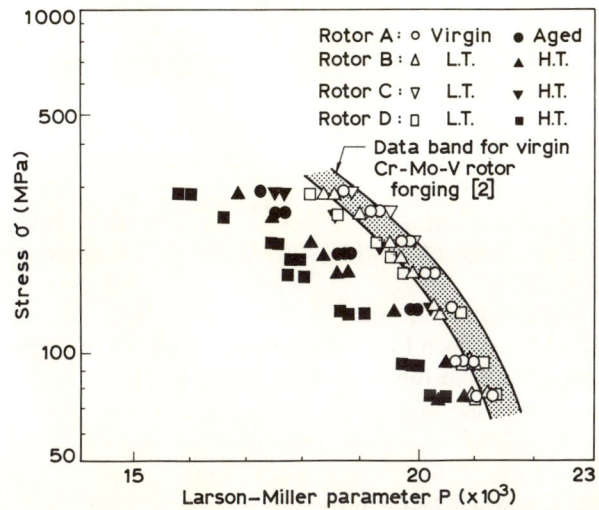

FIG. 5. Creep rupture test results for a Cr–Mo–V rotor forging.

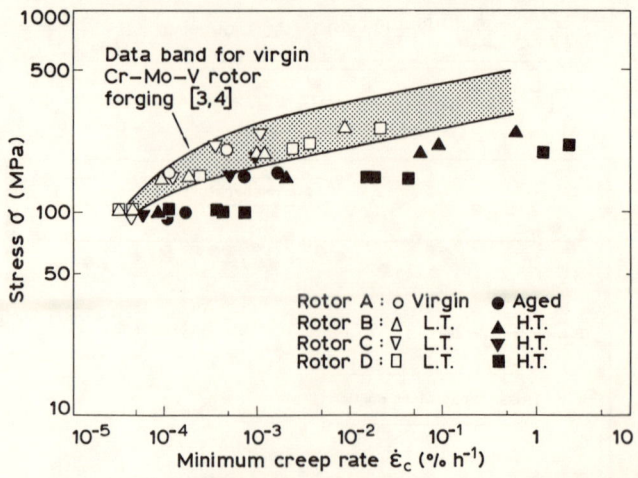

FIG. 6. Creep rate test results for a Cr–Mo–V rotor forging at 566 °C.

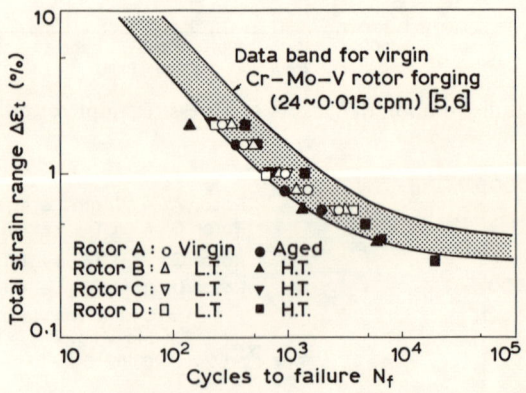

FIG. 7. Low cycle fatigue test results for a Cr–Mo–V rotor forging at 566 °C.

the rotors investigated. The aged rotor and high temperature portion of the retired rotors show a lower strength than the virgin rotor and low temperature portion of the retired rotors.

## DISCUSSION

### Softening in Cr–Mo–V Rotor Forgings

Hardness measurements of the laboratory-aged rotor and of rotors retired after long-term use at elevated temperatures clarified that Cr-

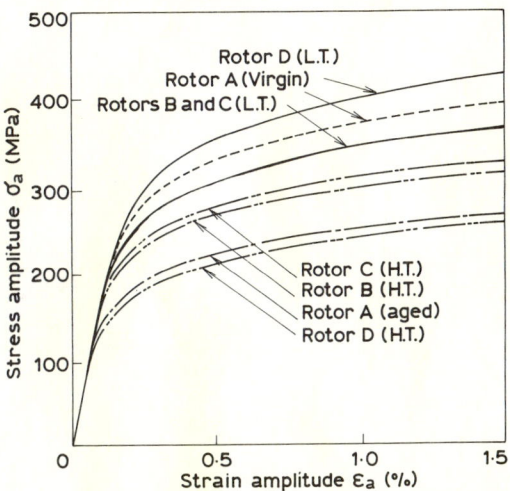

FIG. 8. Cyclic stress-strain curves for a Cr-Mo-V rotor forging at 566 °C.

Mo-V rotor forgings become softer with service, depending on the temperature and time. The hardness data in Fig. 3 are reproduced in Fig. 9, in which the ratio of hardness after softening to the original hardness is plotted against the time–temperature parameter $P'$ expressed as

$$P' = (T + 273)(20 + \log t) \qquad (2)$$

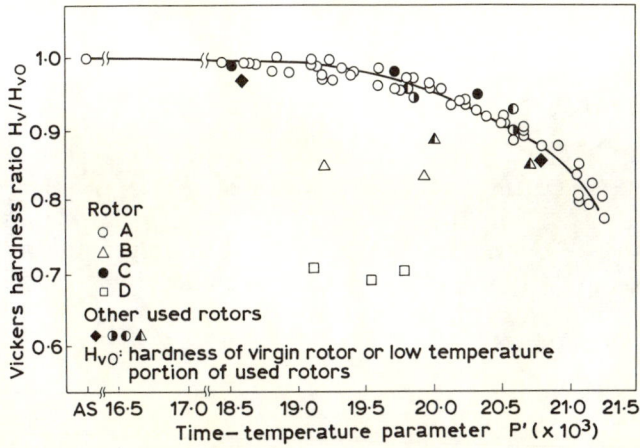

FIG. 9. Time-temperature dependence of hardness for laboratory aged and used Cr-Mo-V rotor forgings.

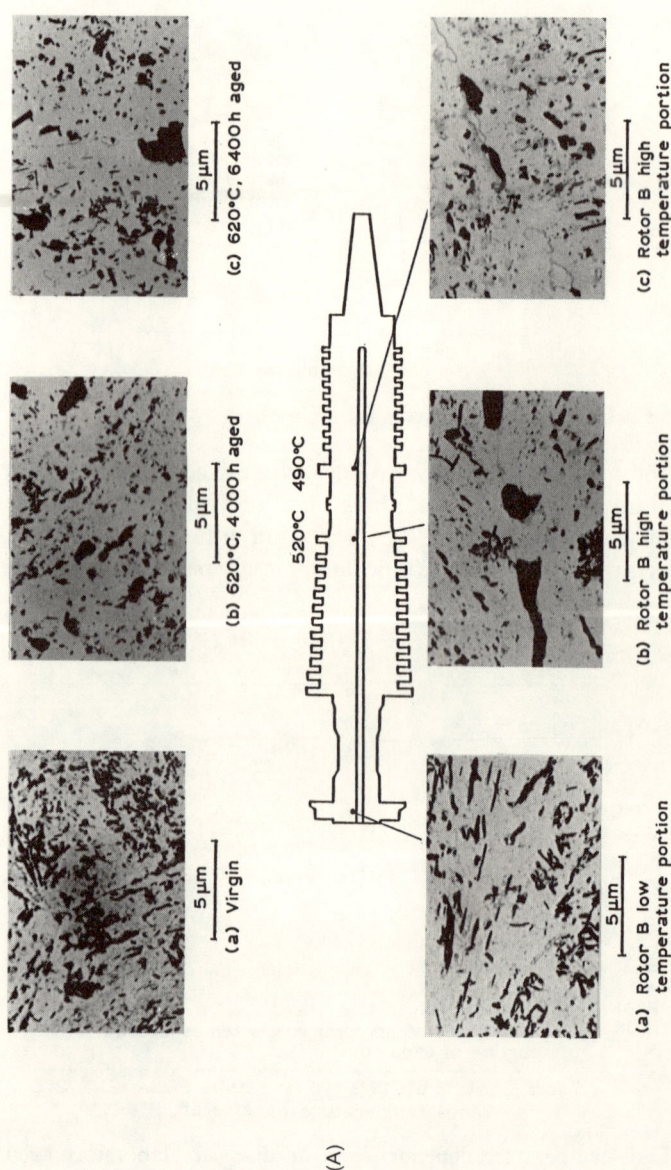

FIG. 10. (A) Carbide microstructure of virgin, aged and retired rotors. (B) Dislocation microstructure of virgin, aged and retired rotors.

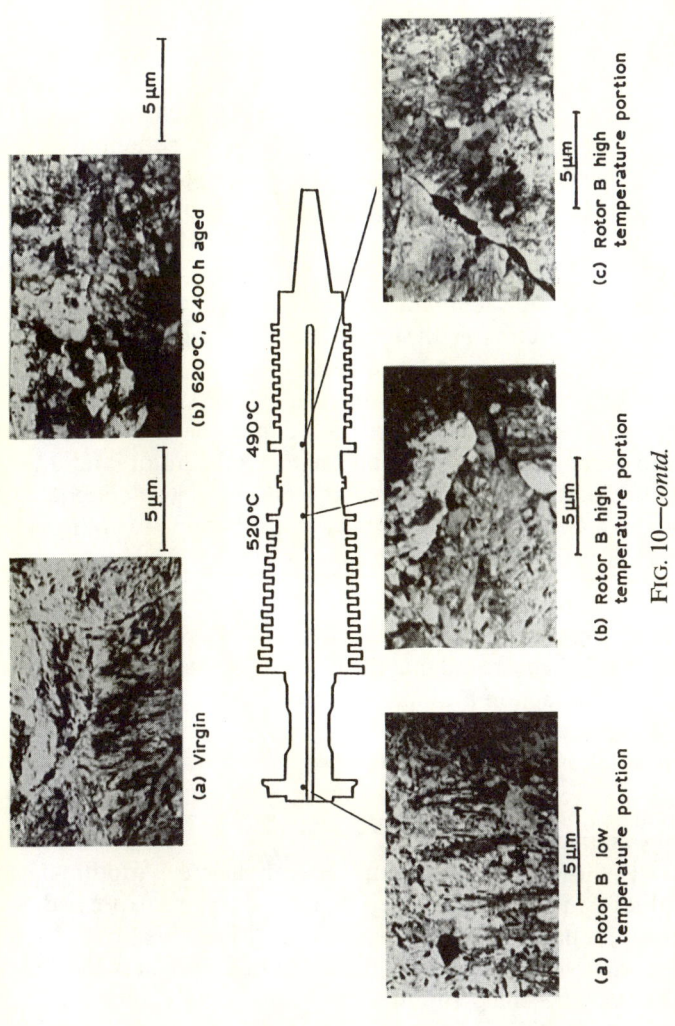

(a) Virgin

(b) 620°C, 6400 h aged

5 μm

5 μm

520°C 490°C

(a) Rotor B low temperature portion

(b) Rotor B high temperature portion

(c) Rotor B high temperature portion

5 μm

5 μm

5 μm

Fig. 10—contd.

(B)

where $T$ is the aging or operating temperature (°C) and $t$ is the aging or operating time (h).

The hardness decrease is uniquely correlated with the time–temperature parameter $P'$, irrespective of the aging temperature for the laboratory-aged rotor samples. Some data for the hardness measured on the retired rotors fit the hardness–parameter relationship for the laboratory-aged rotor, but others do not and show unexpectedly low values. Why this discrepancy appeared is not yet clear.

To learn more about the reason for material softening, a micro-structural investigation was made using a scanning electron microscope and a transmission electron microscope to observe the carbide microstructure and dislocation structure. Figures 10(A) and 10(B) show the results for the virgin and laboratory-aged rotors, and the low temperature and high temperature portions of the retired rotors.

The virgin rotor and the low temperature portion of the retired rotors had almost the same microstructure consisting of finely dispersed carbide precipitates and densely distributed dislocations. On the other hand, carbide coarsening, a dislocation density decrease and subgrain formation were commonly found in the aged rotor and the high temperature portion of the retired rotors. Thus, it is concluded that softening of a Cr–Mo–V rotor forging is attributable to changes in the carbide microstructure and dislocation structure.

### Relationship between Hardness and Mechanical Properties

It has already been stated that the Cr–Mo–V rotor became softer than the original production and that the mechanical properties decreased during long-term exposure to high temperatures. The correlation between hardness and mechanical properties will now be discussed.

*Creep rupture and creep rate*

Figure 11 shows the relationship between the creep rupture data and the hardness, in which the creep rupture data are expressed by the Larson–Miller parameter $P$ shown in Eqn. (1). In this figure, the creep rupture parameter $P$ is linearly correlated to the Vickers hardness $H_v$, depending on the creep rupture stress, and this relationship can be approximated by [7]

$$P = C(\sigma)H_v + D(\sigma) \tag{3}$$

where $C(\sigma)$ and $D(\sigma)$ are functions of stress only. By plotting $C(\sigma)$ and

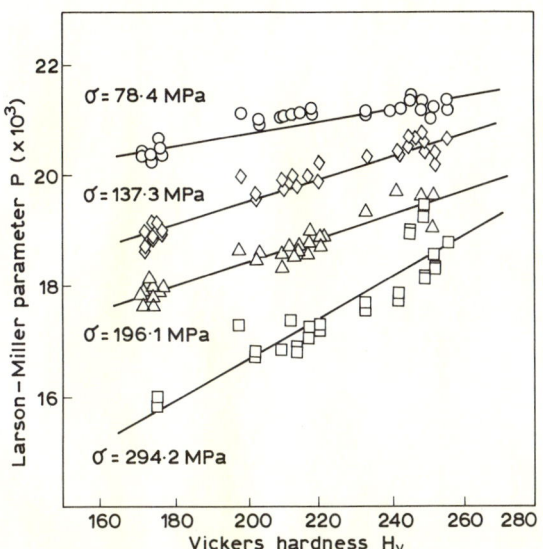

FIG. 11. Relationship between the Larson–Miller parameter and the Vickers hardness for a Cr–Mo–V rotor forging (estimated from Eqn. (3)).

$D(\sigma)$ against the stress $\sigma$ as shown in Fig. 12, the following correlations were obtained:

$$C(\sigma) = C_1 + C_2 \log_{10} \sigma + C_3 (\log_{10} \sigma)^2 \qquad (4a)$$

$$D(\sigma) = D_1 + D_2 \log_{10} \sigma + D_3 (\log_{10} \sigma)^2 \qquad (4b)$$

where $C_i$ and $D_i$ ($i$ = 1, 2 and 3) are material constants.

By combining Eqns. (3) and (4), the creep rupture parameter $P$ can be predicted from the hardness measurement. As shown in Fig. 13, the creep rupture data could be accurately approximated from the hardness by using Eqns. (3) and (4). This correlation is valid even for pre-creep Cr–Mo–V steels [8, 9], as shown in Fig. 14. This suggests that the dominant mechanism for softening during creep is no different from that for thermal aging, i.e. coarsening of the carbide and recovery of the dislocation structure.

Figure 15 shows data plotted for the minimum creep rate $\dot{\varepsilon}_c$ and creep rupture time $t_r$. The well-known Monkman–Grant correlation [10] expressed as below was observed for the rotors tested, irrespective of their hardness:

$$\log_{10} t_r + m \log \dot{\varepsilon}_c = C_0 \qquad (5)$$

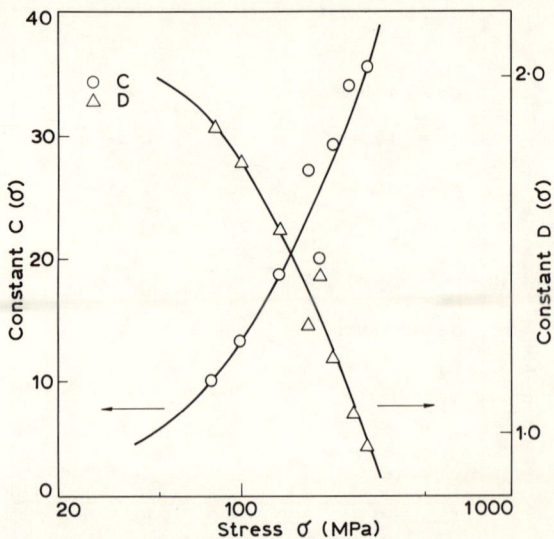

FIG. 12. Stress dependence of constants in the hardness–parameter relationship for a Cr–Mo–V rotor forging.

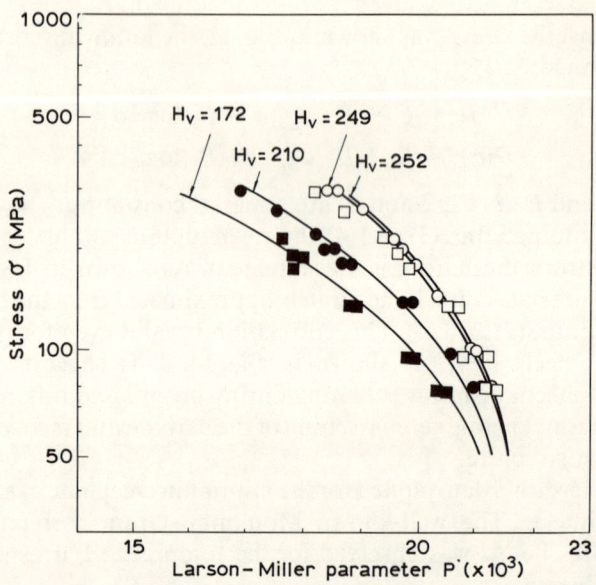

FIG. 13. Estimation of the creep rupture curve by hardness for a Cr–Mo–V rotor forging. Experimental data: $\bigcirc$, $H_v = 252$; $\square$, $H_v = 249$; $\bullet$, $H_v = 210$; $\blacksquare$, $H_v = 172$. ——, Estimated from Eqns. (3) and (4).

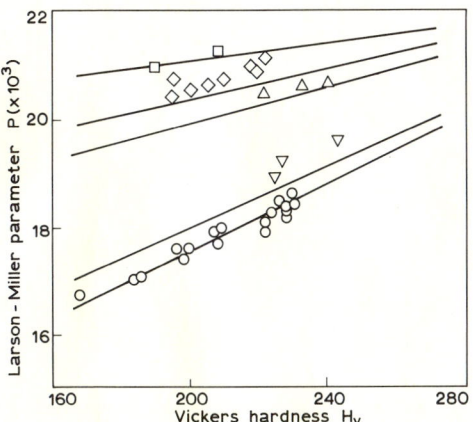

FIG. 14. Relationship between the Larson–Miller parameter and hardness of pre-creep Cr-Mo-V rotor forgings: □, 58·8 MPa [8]; ◊, 98·1 MPa [8]; △, 117·7 MPa [8]; ▽, 215·7 MPa [8]; ○ 241·3 MPa [9]; ——, Estimated from Eqns. (3) and (4).

where $m$ and $C_0$ are material constants. From Eqns. (3)–(5), a correlation between the minimum creep rate $\dot{\varepsilon}_c$ and the hardness $H_v$ can be obtained [7, 11]:

$$\log_{10}\dot{\varepsilon}_c = \frac{1}{m}\left[ C_0 + 20 - \frac{1}{T+273}\{C(\sigma)H_v + D(\sigma)\}\right] \qquad (6)$$

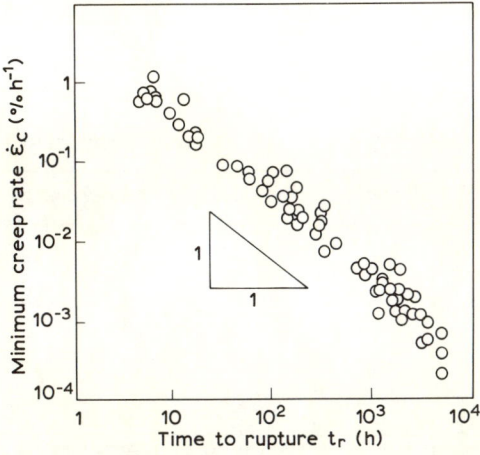

FIG. 15. Monkman–Grant relationship for test samples of a Cr-Mo-V rotor forging.

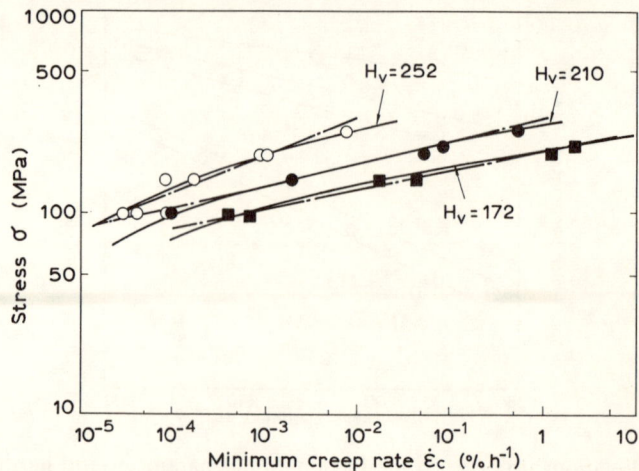

FIG. 16. Estimation of the minimum creep rate by hardness for a Cr–Mo–V rotor forging. Experimental data: $\bigcirc$, $H_v = 252$; $\bullet$, $H_v = 210$; $\blacksquare$, $H_v = 172$. — · —, Norton-type best-fit curve. ———, Estimated from Eqn. (6).

As shown in Fig. 16, Eqn. (6) predicted the experimental data for each sample of different hardness better than the conventional Norton-type best-fit line of Eqn. (7) [12]:

$$\dot{\varepsilon}_c = A_0 \sigma^n \tag{7}$$

where $A_0$ and $n$ are material constants.

The material constants associated with the creep properties in Eqns. (3)–(6) are listed in Table 4.

TABLE 4
Constants associated with creep properties

| Sampling direction | | Axial and circumferential |
| --- | --- | --- |
| Creep rupture | $C_1$ | 71·42 |
| | $C_2$ | −91·43 |
| | $C_3$ | 31·18 |
| | $D_1$ | −9 153·9 |
| | $D_2$ | 38 275·0 |
| | $D_3$ | −12 438·8 |
| Creep | $C_0$ | 0·739 |
| | $m$ | 0·927 |

*Low cycle fatigue properties*

The low cycle fatigue property, the relationship between the applied strain range $\Delta\varepsilon_t$ and number $N_c$ of cycles to crack initiation, is generally expressed by the Manson–Coffin and Basquin life–strain relationsip [13, 14] as follows:

$$\Delta\varepsilon_t = \Delta\varepsilon_e + \Delta\varepsilon_p \tag{8a}$$

$$= k_1 N_c^{\alpha_1} + k_2 N_c^{\alpha_2} \tag{8b}$$

where $\Delta\varepsilon_e$ is the elastic strain range, $\Delta\varepsilon_p$ is the plastic strain range, and $k_1, \alpha_1, k_2$ and $\alpha_2$ are material constants.

From the data shown in Fig. 7, the material constants $k_1, \alpha_1, k_2$ and $\alpha_2$ were obtained by a regression procedure and plotted against the Vickers hardness $H_v$. Figure 17 shows the hardness dependence of the material constants $k_1$ and $\alpha_1$. For all the rotors investigated, $\log k_1$ and $\alpha_1$ were linearly dependent on the material hardness, and their relationships could be approximated as follows:

$$\log k_1 = EH_v + F \tag{9a}$$

$$\alpha_1 = GH_v + H \tag{9b}$$

where $E$, $F$, $G$ and $H$ are constants. On the other hand, the material constants $k_2$ and $\alpha_2$ in the plastic term of Eqn. (8b) were independent of material hardness. Using Eqns. (8) and (9), and the hardness-independent values of $k_2$ and $\alpha_2$, the low cycle fatigue property could be uniquely determined by non-destructive measurements of the material hardness. Examples of the experimental data for low cycle fatigue tests and estimated curves from the hardness measurements are shown in Fig. 18, in which fairly good agreement can be seen.

The cyclic stress–strain relationship for all the rotors investigated, whose material hardness values were different from each other, is shown in Fig. 8. From these curves, the 0·02% offset cyclic yield strengths $\sigma_{yd}$ were obtained and then plotted against the material hardness $H_v$ as shown in Fig. 19. A linear relationship exists between them, and the approximate formula was determined as follows:

$$\sigma_{yd} = IH_v + J \tag{10}$$

where $I$ and $J$ are material constants.

The hardness-dependent and hardness-independent material constants associated with the low cycle fatigue property and cyclic stress–strain relationship are listed in Table 5.

TABLE 5

Constants associated with low cycle fatigue property and
cyclic stress–strain relationship

| Sampling direction | | Axial | Circumferential |
|---|---|---|---|
| Low cycle fatigue | $E$ | | $2 \cdot 59 \times 10^{-3}$ |
| | $F$ | | $-0 \cdot 90$ |
| | $G$ | | $1 \cdot 64 \times 10^{-4}$ |
| | $H$ | | $-0 \cdot 09$ |
| | $k_2$ | $100 \cdot 0$ | $22 \cdot 0$ |
| | $a_2$ | $-0 \cdot 74$ | $-0 \cdot 59$ |
| Cyclic stress–strain | $I$ | | $1 \cdot 07$ |
| | $J$ | | $-40 \cdot 6$ |

The foregoing discussion has indicated the hardness dependence of
the mechanical properties of Cr–Mo–V rotor forgings. The mechanical
properties of creep, creep rupture and fatigue were all dependent on and
quantitatively correlated with the material hardness.

## Creep and Fatigue Life Prediction Method Based on the Non-destructive Measurement of Hardness

Because of its simplicity, a life prediction is usually made by the
cumulative damage rule [15], in which the accumulation of creep
damage $\phi_c$ and the fatigue damage $\phi_f$ are expressed by the time and cycle
fractions as follows:

$$\phi_c = \int \frac{dt_i}{t_{ri}} \tag{11a}$$

$$\phi_f = \sum \frac{n_j}{N_{cj}} \tag{11b}$$

where $t_{ri}$ is the creep rupture time under the loading condition $i$, $t_i$ is the
duration of time under the loading condition $i$, $N_{cj}$ is the number of
cycles to crack initiation under the loading condition $j$ and $n_j$ is the
number of applied cycles under the loading condition $j$.

It has already been stated that the creep and fatigue properties of
Cr–Mo–V forgings were degraded during long-term service at high
temperatures, and that these properties were functions of the material
hardness, which can be non-destructively measured on the actual
components. Therefore, creep damage and fatigue damage are functions
of hardness as well as the duration of time and number of applied

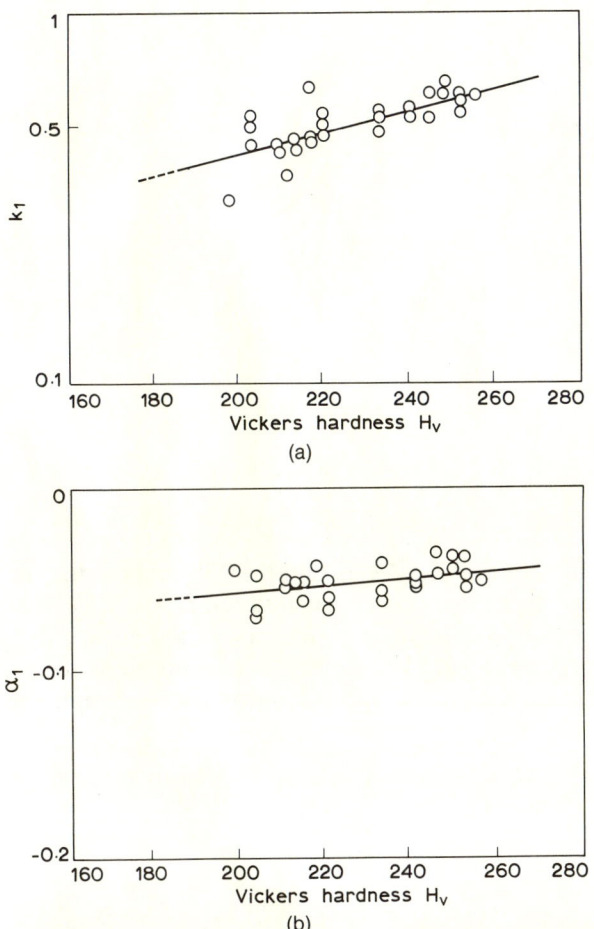

FIG. 17. Relationship between the hardness and low cycle fatigue property for a Cr–Mo–V rotor forging at 566 °C. (a) Hardness versus $k_1$; ——, estimated from Eqn. (9a). (b) Hardness versus $\alpha_1$; ——, estimated from Eqn. (9b).

cycles, and the hardness effect on damage must be taken into account when the crack initiation life of a Cr–Mo–V rotor is estimated according to Eqn. (11) [16].

Figure 20 shows the fatigue life assessment procedure for a groove on the outer surface of a Cr–Mo–V steel steam turbine rotor. The hardness of the rotor is non-destructively measured first, in order to reflect the material degradation (softening) and its effect on fatigue damage and residual life evaluation.

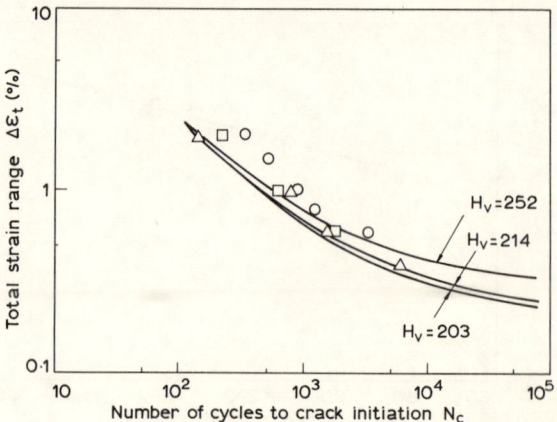

FIG. 18. Estimation of low cycle fatigue properties by hardness for a Cr–Mo–V rotor forging at 566 °C; ○, $H_v$ = 252; △, $H_v$ = 214; □, $H_v$ = 203 (estimated from Eqns. (8) and (9)).

The local strain range $\Delta\varepsilon_t$ is obtained by multiplying the nominal strain range $\Delta\varepsilon$ by the strain concentration factor $K_\varepsilon$. $K_\varepsilon$ is determined from the relationship between $K_\varepsilon$ and $\lambda$, the former having been established for the given elastic stress concentration factor $K_t$ and the latter being the ratio of the applied nominal stress amplitude $\Delta\sigma_0/2$ to

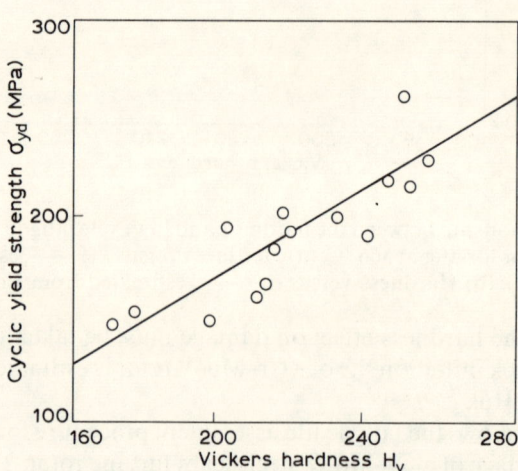

FIG. 19. Relationship between the 0·02% cyclic yield strength and hardness for a Cr–Mo–V rotor forging at 566 °C. ——, Estimated from Eqn. (10).

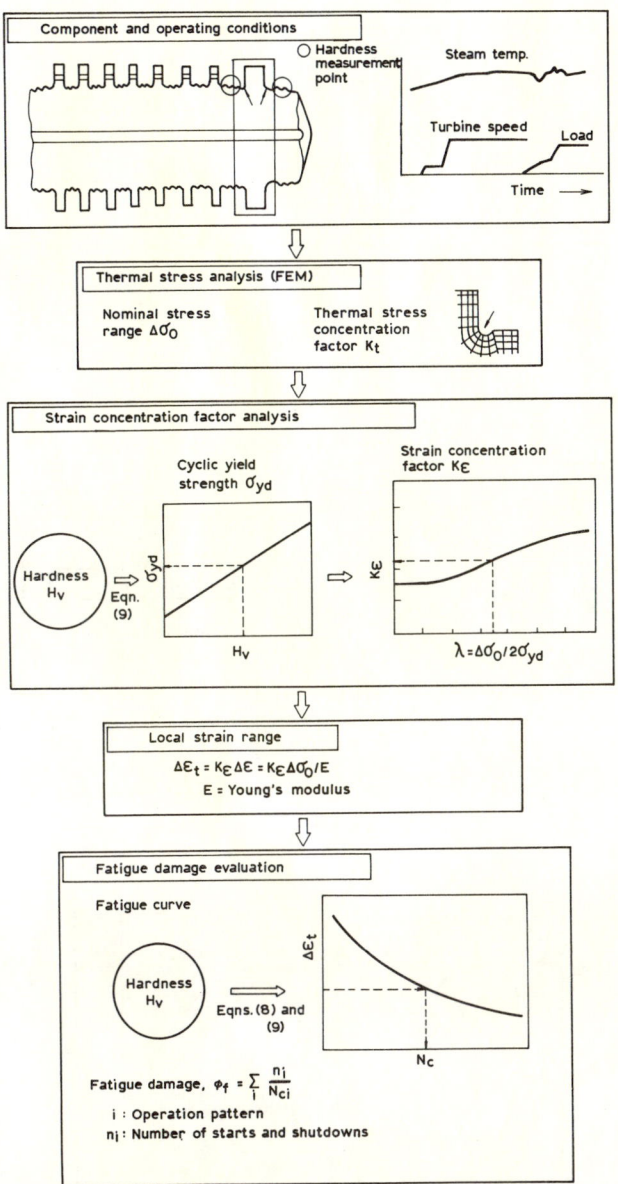

FIG. 20. Fatigue damage estimation procedure.

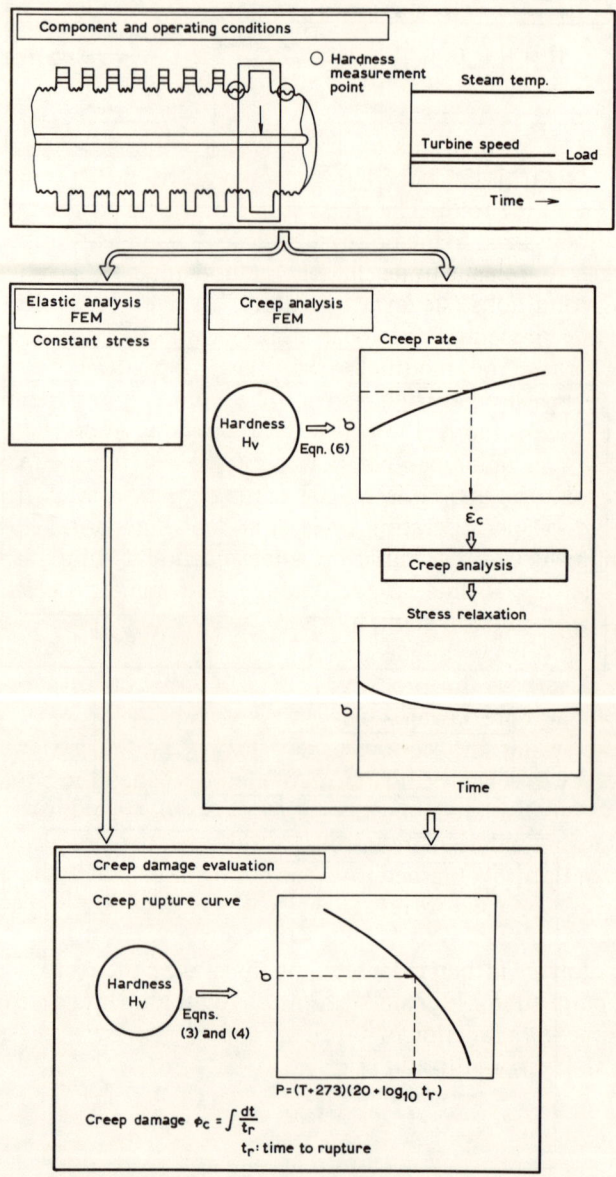

FIG. 21. Creep damage estimation procedure.

the cyclic yield strength $\sigma_{yd}$ [17]. The applied nominal stress, $\sigma_0$ can be calculated from

$$\sigma_0 = \frac{E\alpha}{1-v}(T_{av} - T_s) \tag{12}$$

where $E$ is Young's modulus, $\alpha$ is the modulus of thermal expansion, $v$ is Poisson's ratio, $T_s$ is the rotor surface temperature and $T_{av}$ is the average temperature of the rotor. The stress concentration $K_t$ is the ratio of the maximum stress $\sigma_{max}$ to the nominal stress $\sigma_0$ at the rotor surface groove calculated by the elastic finite element method under the given transient operating conditions (steam temperature, turbine speed, electric load etc.). In this procedure for local strain evaluation, the cyclic yield strength $\sigma_{yd}$ can be modified from the information on the non-destructive measurement of hardness. The fatigue damage per transient operating cycle is the reciprocal of the number $N_c$ of cycles to fatigue crack initiation, which can be obtained by comparing the local strain range $\Delta\varepsilon_t$ and the low cycle fatigue curve expressed as Eqn. (8). The fatigue damage per operating cycle is added in accordance with the operating history to determine the total accumulated fatigue damage and residual life. In this fatigue damage assessment procedure, non-destructive information about hardness measurements is introduced to modify the low cycle fatigue curve.

Figure 21 shows the procedure for estimating creep damage in a Cr–Mo–V rotor bore during steady-state operation. The bore temperature and stress under the condition of steady-state operation can be calculated by the elastic finite element method. When the bore stress is high and it is considered that stress relaxation is not negligible, the finite element creep analysis is required. In this case, the creep constitutive equation of Eqn. (6) obtained from the non-destructive measurement of hardness can be used to reflect the material degradation (softening) effect.

Creep damage is then calculated by integrating the reciprocal of the creep rupture time $t_r$ during steady-state operation. Again, the creep rupture time $t_r$ is a function of hardness, and the hardness measurement data are thus fully utilized.

## CONCLUSION

Laboratory-aged Cr–Mo–V rotors and Cr–Mo–V rotors retired after 130 000–140 000 h of service at elevated temperatures were cut into specimens and the material properties were investigated. The results obtained are listed below.

(1) The high temperature portion of a Cr–Mo–V steam turbine rotor degrades during long-term service at high temperatures.

(2) Typical degradation is softening, which is caused by the coarsening of carbide and the recovery of dislocation structure.

(3) Quantitative relationships between the hardness and mechanical properties were established for creep rate, creep rupture time, low cycle fatigue and the cyclic stress–strain relationship.

(4) A new procedure for creep and fatigue life assessment was established, which is based on non-destructively obtained information on the material degradation, i.e. softening. Its application to the assessment of creep and fatigue life in a Cr–Mo–V steam turbine rotor was demonstrated.

## REFERENCES

[1]  C. E. Jaske, H. Mindlin and J. S. Perrin, *ASTM STP 519* (1973).
[2]  *NRIM Creep Data Sheet No. 9A* (1979).
[3]  R. M. Goldhoff and G. J. Hahn, *ASM Publication D8-100* (1969).
[4]  *Data Book for High-Temperature Strength of Metals*, 1, The Iron and Steel Institute of Japan (1972) (in Japanese).
[5]  K. Kuwabara and A. Nitta, *CRIEPI Report 277008* (1977) (in Japanese).
[6]  M. M. Levin, *Exp. Mechanics*, 13 (9), 353 (1973).
[7]  K. Kimura, K. Fujiyama, F. Ishii and M. Muramatsu, *Int. Conf. on Creep* (1986).
[8]  M. Yamada, O. Watanabe, S. Komatsu and S. Nakamura, *Report of the 123rd Committee, Japan Society for the Promotion of Science*, 22 (1), (1981) (in Japanese).
[9]  R. M. Goldhoff and D. A. Woodford, *ASTM STP 515* (1972).
[10]  F. C. Monkman, *ASTM Preprint 72* (1956).
[11]  K. Kimura, T. Inukai, K. Fukuda and M. Muramatsu, *Int. Conf. on Creep* (1986).
[12]  C. F. Etienne, H. C. Van Helst and P. M. Eijers, *Creep of Engineering Materials and Structures*, Applied Science, London (1978).
[13]  S. S. Manson, *NACA TN 2933* (1953).
[14]  L. F. Coffin, Jr., *Proc. 4th Sagamore Conf.*, 219 (1957).
[15]  *ASME Boiler and Pressure Vessel Code, Case N-47*, ASME.
[16]  K. Kimura, K. Fujiyama, M. Muramatsu and M. Akiba, *Proc. Int. Conf. on Life Extension and Assessment of Fossil Plants, Washington* (1986).
[17]  D. C. Gonyea, *ASTM STP 520* (1973).

# Index

A 286 steel
  fatigue crack propagation in, 81
  thermal–mechanical fatigue life
    characteristics, 207, 213
A 387 steel
  fatigue life data, 188
  material constants used in life
    prediction, 197
  welded joints with stainless steels
    fatigue life data, 188–92
    hardness distribution, 187–8
    life prediction, 196–201
    low cycle fatigue fracture
      behavior, 194–5
    strain distribution, 192–4
    welding conditions, 186
  see also Chromium–molybdenum
    steels, $2\frac{1}{4}$Cr–1Mo steel
Accelerated testing procedures, 137,
  138–41, 162
Air, creep–fatigue properties affected
  by
  austenitic stainless steels, 141–7
  chromium–molybdenum steels,
    147–52
AISI 316 steel. See Stainless steels,
  type 316
AISI 405 steel. See Stainless steels,
  type 405
Anisotropic damage theory, 56–7, 59

ASME Boiler and Pressure Vessel Code
  life estimation methods used, 34,
    38, 223
  welded joint design, 184
Austenitic stainless steels
  chemical composition of, 118
  crack growth/closure behavior,
    73–4, 79, 94, 96–100, 104–5
  creep fracture modes of, 120–1, 125
  creep–fatigue interaction, 122–5
  creep–fatigue loading tests, 121–2
  environmental effects on creep-
    fatigue properties, 141–7
  fatigue test data, 128–9
  life prediction for, 87, 88, 129–33
  stress–strain curves, 10, 12
  yield locii for, 17
  see also Stainless steels

Basquin life–strain relationship, 263
Bauschinger effect, 17

$C^*$-integral, 92
  see also $J^*$-integral
Carbon steels
  creep crack propagation in, 68–9,
    71, 72, 74–7, 83
  fatigue crack propagation in, 79–81

271